Omics Technologies and Crop Improvement

Omics Technologies and Crop Improvement

Editor

Mr. Ranveer Singh

scitus
academics

Omics Technologies and Crop Improvement

Edited by **Ranveer Singh**

ISBN: 978-1-68117-018-3

Library of Congress Control Number: 2015931821

SCITUS Academics LLC,
www.scitusacademics.com
Box No. 4766, 616 Corporate Way,
Suite 2, Valley Cottage,
NY 10989

Printed in United States of America on Acid Free Paper ∞

Contents

Preface

Given the ever-increasing population and the decrease in natural resources, food production needs to be improved, which can perhaps be achieved by improving the quantitative and qualitative characteristics of crop plants through the adoption of new analytical instruments and technologies. In the last ten years or perhaps so rapid progress in plant biology has been made, especially with the introduction of high throughput omics technology. The three main omics technologies–genomics, proteomics and metabolomics–involve characterizing and quantifying genomes, metabolome and proteomes with extremely rapid, miniaturized and automated methods respectively. In fact, gene expression and gene discovery profiling technologies create an unprecedented opportunity for plant breeders who can now use molecular markers to evaluate and improve diversity in their germplasm collections, to introduce valuable traits from new sources and to identify genes that control key characteristics. Genomics technologies are used to decipher the multigenicity of abiotic and biotic plant stress reactions through genome sequences, stress-specific tissue and cell transcript collections, transcripts, protein and metabolite profiles and dynamic changes in their interactions, mutant screens and protein. As a result of the use of high-performance omics methods, large amounts of raw data are generated and stored, processed and analyzed using bioinformatics tools.

This book focuses on how omics can help crop science and horticulture and shows the advantages of using these technologies to increase other features and crop yields, such as nutritional values and strength. The book also highlights crop improvements such as higher yields, disease resistance, value-added performance and non-transgenic resistance.

Chapter 1

Mapping Epistasis and Environment × Qtx Interaction Based on Four -Omics Genotypes for the Detected Qtx Loci Controlling Complex Traits in Tobacco

Liyuan Zhou[a, 1], Ruiyuan Li[b, 1], Longjiang Fan[a], Yuewei Shi[b], Zhihong Wang[b], Shengdon Xie[b], Yijie Gui[a], Xueliang Ren[b,] ,and Jun Zhu[a,]

[a]Key Laboratory of Crop Germplasm Resource of Zhejiang Province, Department of Agronomy, Zhejiang University, Hangzhou, Zhejiang 310058, China.

[b]Key Laboratory of Molecular Biology, Guizhou Academy of Tobacco Sciences, Guiyang, Guizhou 550013, China.

ABSTRACT

Using newly developed methods and software, association mapping was conducted for chromium content and total sugar in tobacco leaf, based on four -omics datasets. Our objective was to collect data on genotype and phenotype for 60 leaf samples at four developmental stages, from three plant architectural positions and for three cultivars that were grown in two locations. Association mapping was conducted to detect genetic variants at quantitative trait SNP (QTS) loci, quantitative trait transcript (QTT) differences, quantitative trait protein (QTP) variability, and quantitative trait metabolite (QTM) changes, which can be summarized as QTX locus variation. The total heritabilities of the four -omics loci for both traits tested were 23.60% for epistasis and 15.26% for treatment interaction. Epistasis and environment × treatment interaction had important impacts on complex traits at all -omics levels. For decreasing chromium content and increasing total sugar in tobacco leaf, six methylated loci can be directly used for marker-assisted selection, and expression of ten QTTs, seven QTPs and six QTMs can be modified by selection or cultivation.

INTRODUCTION

Most important agronomic traits are complex [1]. Decoding the genetic constitution of complex traits and obtaining information on phenotypic variation are some of the most important challenges of genetic analysis. In contrast to Mendelian traits controlled by individual major genes, the phenotypic variations of complex traits are due to segregation of multiple loci with small effects which are sensitive to environmental factors. Using gel-based or next generation sequencing and molecular marker analysis technology, genetic linkage analysis of quantitative trait locus (QTL) has become one of the most commonly used techniques in complex trait analysis [2] and [3]. QTL analysis can also be combined with available transcript, protein and metabolite profiles for a mapping or association population generally resulting in regression analysis between markers and endogenous phenotypes (e.g. gene expression levels, protein modification, or levels of a particular secondary metabolite). By using such molecular, protein or biochemical variants as trait phenotypes, the linkage or association QTL mapping is known as expression-QTL (eQTL), protein-QTL (pQTL) and metabolite-QTL (mQTL), respectively.

These full pathway molecular phenotypes, from transcript to translated protein to metabolic product, help elucidate genotypic variation that underlies morphological and physiological traits [4]. However, due to the limited recombination events in the mapping population derived from bi-parental crosses, regardless of the choice of either molecular variants or complex phenotypic traits, the QTLs detected via linkage analysis can only be mapped to large genomic regions [5]. Recently, the increasing use of high-throughput molecular techniques from the -omics sciences (genomics, transcriptomics, proteomics and metabolomics) has created a huge amount of -omics data, which can be applied to traditional genetic or agronomic experiments [6].

Recent genotyping methods (e.g. for SNPs) produce much higher density genome coverage than traditional gel-based molecular markers, and thus are much more informative. This leads to the potential of association mapping for complex trait analyses [7]. Compared with linkage mapping, association mapping is a high-resolution method based on linkage disequilibrium (LD), and has recently been applied to plant populations [8], [9] and [10]. Here we propose that instead of just using SNPs as variants in LD analysis for the detection of QTL, the molecular variants of four -omics datasets can also be used as generalized genotypes in association mapping for complex traits. This multi-omics approach would be crucial for the identification of what we term quantitative trait SNPs (QTS), quantitative trait transcripts (QTT) [5], quantitative trait proteins (QTP), and quantitative trait metabolites (QTM). The association mapping based on the four -omics datasets can in compendium or in conjunction be called QTX mapping, a more general term we suggest for use in this type of research.

In addition to the detection of QTX themselves, G × G interaction (epistasis) and G × E interaction can also be detected by QTX mapping. These interaction effects may explain a considerable proportion of the missing heritability associated with QTL based on individual molecular marker loci [11]. In general, the size of datasets involved in QTX mapping will be an order of magnitude larger than the size of datasets for typical QTL detection. This has presented a challenge that has hardly been matched by contemporary hardware, making

QTX analysis difficult to perform efficiently until recently with the application of GPU (Graphic Processor Unit) parallel computation which has significantly increased the ability to solve computationally intensive biological problems [12] and [13]. GPU parallel computation addresses the ever-increasing demand for higher computational speed and has paved the way for the analysis of -omics data from large scale or multiple layer experiments.

Tobacco (*Nicotiana tabacum* L.) is one of the most important model plants in genetic analysis. The quality of tobacco leaves is determined by the composition and quantity of metabolites [14], which are quantitative traits controlled by multiple genes and environmental factors. Previous studies on genetic architecture and regulated network of such complex traits were unable to comprehensively dissect the mechanism of catabolism, anabolism or accumulation of these metabolites [15], [16], [17], [18], [19], [20], [21] and [22]. Implementation of QTX mapping by using various types of -omics datasets in tobacco was predicted as a useful opportunity to illustrate the regulated networks involved in genetic control of these complex traits. Therefore, for this study, we conducted QTX mapping to reveal the genetic architecture of two complex traits in tobacco leaves. Four -omics datasets were used as genotypes: genome methylation data for QTS, transcription and miRNA levels for QTT, enzyme and protein amounts for QTP, and metabolite levels for QTM. The genotypic and phenotypic data were collected from 60 tobacco leaf samples of three cultivars, from four development stages and three positions in the plants and grown in two cultivation environments. The potential application of the QTX results in breeding practice was also discussed.

MATERIALS AND METHODS

Plant Growth and Sample Preparation

In 2010, three tobacco cultivars (K326, Hongda and Zunyan 6) were grown in farm fields at Guiding (26.58° N, 107.23° E) and Xingyi (25.08° N, 104.90° E) under normal condition for crop production in Guizhou province, China. The plants of three cultivars were planted in 10 rows with 30 plants per row and in three blocks. The plant-to-plant spaces between and within rows were 100 cm and 50 cm, respectively. For each of the three cultivars, leaves of 25 plants from 5 points were pooled for 1) the combination of upper or middle leaves from two locations within the plant at four developmental time points with sampling every 12 days, and 2) the lower leaves from two locations within the plant at two developmental time points, thus resulting in a total of 60 samples. The pooled leaves were immediately frozen in liquid nitrogen and stored at – 80 °C for further use.

Methylation Profiling and Clone Sequencing

A methylation DArT chip for tobacco was developed by Diversity Array Technology Ltd. (Canberra, Australia) as described at http://www.diversityarrays.com/dnamethylation.html. Total DNA was extracted and hybridization followed the DArT methylation profiling protocol as described by Lu et al. [23]. The program DArT Soft was used to determine whether the fragments in the arrays tested for each sample were methylated or not.

mRNA Array Design, Hybridization and Normalization

A custom-designed microarray platform was used for the analysis of total RNA extracted from the tobacco leaf samples. The microarray was comprised of three 60-mer probes for each of 44,873 unigenes derived from public Expressed Sequence Tags (ESTs) of tobacco and was made following a protocol provided by Roche Co. (http://www.nimblegen.com/). The 60-mer probes were chosen from a group of six to seven non-overlapping probes designed against different parts of each gene model. The probes with E-values most similar to the average of the six to seven non-overlapping experimental probes were assumed to be the most reliable for transcript level estimation. Total RNA was extracted with the RNeasy Mini Kit (Qiagen Corp, Valencia, CA, United States) and DNase treated in-column with the RNase-Free DNase set (also from Qiagen). Double-stranded cDNA was synthesized using the SuperScript Double Strand cDNA Synthesis Kit (Invitrogen Inc., Carlsbad, CA, United States) with oligo (dT) primers following the manufacturer's protocol. Cy-3 and Cy-5 labeling and hybridization steps were performed by NimbleGen using standard procedures (http://www.nimblegen.com/). Expression values were generated by Roche NimbleGen proprietary software using quantile normalization [24] and the Robust Multichip Average algorithm [25].

miRNA Array Design, Hybridization and Normalization

A total of 1000 candidate tobacco miRNA genes were obtained from Guo et al. [26], which were identified by sequencing and advanced bioinformatics analysis of small fragment RNAs. These miRNAs were used to design the miRNA array based on Agilent miRNA chip technology. Total RNA was extracted using mirVanamiRNA Isolation Kit (Applied Biosystems/Ambion, Austin, TX, United States), and RNA concentrations were determined with a NanoDrop ND-1000 spectrophotometer (NanoDrop Technologies, Wilmington, DE, United States). Following this, a total of 120 ng of total RNA was fluorescently labeled with Cyanine 3-pCp, and hybridized onto the arrays for 18–20 h at 55 °C. Slides were scanned by an Agilent microarray scanner G2565BA and the images obtained were processed with Feature Extraction Software 9.5.3.1 (also from Agilent). Intensity values were processed using Cluster 3.0 software whereby data were normalized, log transformed, and median centered [27]. Only normalized miRNAs with less than 20% missing values across the samples were included in the subsequent analyses.

Metabolite quantified

Content of Threonine, Lysine, Serine and Phenylalanine was quantified by HPLC (Waters 2695, Waters Alliance). Briefly, 1.0 g dry leaf powder was placed in 50 mL Erlenmeyer flask after sifting with a 40 mm mesh sieve. Totals of 200 μL of 0.1 mg mL^{-1} internal standard solution and 50 mL of ultrapure water were added, and then ultrasonic vibration was conducted for 60 min at room temperature. The resulting suspension was filtered through a 0.45 μm membrane filter. Subsequently, 50 μL of the filtrate was added to a hydrolysis tube, where it was combined with 70 μL AccQ-1 derivatization buffer solution. A shock treatment of 10 s of vigorous stirring using a vortex followed while 20 μL AccQ-2A amino acid derivatization reagent was added. An additional 10 s of shaking was needed after the first vortexing was finished. The extract was then placed in an oven for the full derivatization reaction at 55 °C for 10 min. The solution was then used for HPLC analysis.

Totalsugarandfructosecontentwasquantifiedspectrophotometrically with a Dionex ICS-2000 + ED40. The fresh sample was ground in liquid nitrogen. An aliquot of 0.5 g of ground powder for each sample was then placed into 100-mL volumetric flasks each with 70 mL of deionized water added. Extraction by ultrasound was used for 1 h. The volume was set to the 100-mL mark and separated for 15 min under centrifugation at 9000 r min^{-1}. The supernatant was filtered using a membrane of 0.45 μm pore size (Tianjin Jinteng Experiment Equipment Co., Tianjin, China) to remove impurities, and then passed over a RP pre-treatment column to remove pigments and macromolecules. Finally 0.20 mL of the filtered liquid was taken, diluted to 10.0 mL, and passed through a second membrane of 0.22 μm pore size (Tianjin Jinteng Experiment Equipment Co., Tianjin, China), which the resulting effluent was analyzed. Peak area was quantified by software accompanied with the equipment.

Linolenic acid and linoleic acid were quantified with a gas chromatography (GC) method using an HP-6890 analyzer (Agilent Corp., Santa Clara, CA, United States) following the protocol provided by the vendor. Briefly, 1000 g mixed sample was taken in 100 mL conical flask, then 500 μL of an adipic acid methanol internal standard solution was added along with 25 mL of 10% H_2SO_4–CH_3OH solution. This mixture was shaken by mechanically oscillated overnight at low-speed for the derivatization reaction. The solution was then transferred into 250 mL pyriform separatory funnels with 50 mL distilled water added. The solution was extracted three times by gently shaking with 15 mL CH_2Cl_2, followed by collecting and placing the extract in a 100 mL conical flask with grinding stopper. The appropriate amount of anhydrous sodium sulfate was then added to remove trace water, and the clear and transparent extract was used for analysis.

Chlorogenic acid was quantified by high-pressure liquid chromatography (HPLC) using the LC-2010AHT from Shimadzu Corp. (Shimadzu Corp., Nakagyo-ku, Kyoto, Japan) and default protocol. Briefly, fresh leaf sample was ground in liquid nitrogen and a 0.5 g milled sample taken to a 5 mL centrifuge tube, where 1.5 mL of a

50% aqueous methanol solution was added before treating with ultrasound for 20 min at 56 kHz. The extract was then filtered with liquid membranes (0.22 μm) and stored in a bottle for further analysis.

Chromium content was quantified using microwave digestion and inductively coupled plasma optical emission spectrometry (ICP-OES). A 0.5 g sample was placed in the inner digestion tank of poly-tetra-fluoro-ethylene, which was itself put into an outer tank to which 4 mL of nitrate, 1 mL of hydrogen peroxide, and 0.5 mL of hydrofluoric were added. The sample was sequentially digested by the following procedures in the microwave workstation: digestion at 100 °C for 10 min, at 180 °C for 10 min, and at 220 °C for 20 min. When the digestion was completed, the tank was cooled down to room temperature, and the pressure was reduced to lower than 0.1 MPa. Then the digestion mixture was transferred into a 25 mL volumetric flask after adding 5 mL of boric acid solution, and the inner tank was washed with a small amount of ultrapure water several times, during which the cleaning liquid was merged into the digestion mixture until the final volume was topped up to the original volume. A blank test was performed simultaneously. The parameters of ICP-OES analysis were set as: RF generator transmission power of 1.2 kW; plasma gas flow of 15 L min^{-1}; auxiliary gas flow of 1.5 L min^{-1}; nebulizer pressure of 240 kPa; and cleaning time of 20 s. Measurements were conducted 3 times at intervals of 10 s each. Meanwhile, the peristaltic pump speed was 15 r min^{-1} and a Fitted Model was used to correct for background.

Mixed Linear Model

We used the four -omics datasets to conduct QTX mapping. In order to predict genetic effects of QTXs along with epistasis and QTX × environment interaction, a mixed linear model approach was used by including treatments (three varieties in two locations) as fixed effects (*E*), compared to QTX (*Q* and *QQ*) and QTX × environment interaction (*QE* and *QQE*) as random effects, namely:

$$y = \mu + E + Q + QQ + QQE + \varepsilon \qquad (1)$$

where, y = phenotypic value; μ = population mean; E = total treatment effect; Q = total individual loci effect; QQ = total epistasis effect between loci; *QE* and *QQE* = the total *Q* × environment interaction effect and total *QQ* × environment interaction effect, respectively; = residual effect.

The genetic model for the phenotypic value of the *k*-th genotypes in the *h*-th treatment (y_{hk}) can be expressed by the following mixed linear model,

$$y_{hk} = \mu\_e_h + \sum_i q_i u_{ik} + \sum_{i<j} qq_{ij} u_{ik} u_{jk} + \sum_i qe_{hi} u_{hik}$$
$$+ \sum_{i<j} qqe_{hij} u_{hik} u_{hjk} + \varepsilon_{hk} \qquad (2)$$

where, μ = the population mean; e_h = the fixed effect of the *h*-th treatment; q_i = the *i*-th locus effect with coefficient u_{ik}; qq_{ij} = the epistasis effect of the *i*-th locus by *j*-th locus with coefficient $u_{ik}u_{jk}$; qe_{hi} = the locus × treatment interaction effect of the *i*-th locus in the *h*-th treatment with coefficient u_{hik}; qqe_{hji} = the epistasis × treatment interaction effect of the *i*-th locus and *j*-th locus in the *h*-th treatment with coefficient $u_{hik}u_{hjk}$; and $_{hk}$ = the random residual effect of the *k*-th breeding line in the *h*-th treatment.

The mixed linear model can be presented in matrix notation,

$$y = Xb + U_Q e_Q + U_{QQ} e_{QQ} + U_{QE} e_{QE} + U_{QQE} e_{QQE} + e_\varepsilon$$
$$= Xb + \sum_{v=1}^{4} U_v e_v + e_\varepsilon \qquad (3)$$

where **y** is an $n \times 1$ column vector of phenotypic values and *n* is the sample size of observations; **b** is a column vector of μ, treatments in the experiment; **X** is the known incidence matrix relating to the fixed effects; **U** is the known coefficient matrix relating to the *v*-th random vector **e***v*; **e** ~ *MVN*(**0**, **I** ^{2}) is an$n \times 1$ column vector of residual effects.

Prediction of genetic effects

The estimation of fixed effects (e) and prediction of random effects (*q*, *qq*, *qe* and *qqe*) were obtained using QTXNetwork software based on GPU parallel computation (http://ibi.zju.edu.cn/software/QTXNetwork/). By using mixed linear model approaches described in QTLNetwork 2.0 [28], association was conducted for complex traits against a panel of genetic markers for the QTS dataset, or quantitative expression of transcripts/proteins/metabolites for the QTT/P/M datasets, respectively.

Estimation of Heritability

The total phenotypic variance was considered as the sum of genotype variance ($V_G = V_Q + V_{QQ}$), genotype × treatment interaction variance ($V_{GE} = V_{QE} + V_{QQE}$), and residual variance (V):

$$
\begin{aligned}
V_p &= V_G + V_{GE} + V_\varepsilon \\
&= (V_Q + V_{QQ}) + (V_{QE} + V_{QQE}) + V_\varepsilon \\
&= \left(\frac{1}{df_Q} \sum_i qi^2 + \frac{1}{df_{QQ}} \sum_{i<j} qq_{ij}^2 \right) \\
&+ \left(\frac{1}{df_{QE}} \sum_i qe_{hi}^2 + \frac{1}{df_{QQE}} \sum_{i<j} qqe_{hij}^2 \right) + V\varepsilon.
\end{aligned}
\tag{4}
$$

Individual heritability of specific genetic effects was estimated as $hg^2 = g^2/VP$. Total heritability was defined as the relative contribution of genetic variance to phenotypic variance with the following estimation model:

$$
\begin{aligned}
h^2_{G+GE} &= h^2_Q + h^2_{QQ} + h^2_{QE} + h^2_{QQE} \\
&= \sum h^2_q + \sum h^2_{qq} + \sum h^2_{qe} + \sum h^2_{qqe}
\end{aligned}
\tag{5}
$$

where $h_{G+}GE^2$ = total heritability; hQ^2 = heritability contributed by sum of individual Q_{TXs}, $h_{QQ}{}^2$ = heritability contributed by sum of pair-wise epistasis (qq), hQE^2 = treatment-specific heritability contributed by sum of individual treatment interaction (qe), $h_{QQE}{}^2$ = treatment-specific epistasis heritability contributed by sum of pair-wise epistasis by treatment interaction (qqe); hq^2 = heritability of individual locus effect, hqq^2 = heritability of individual epistasis effect, hqe^2 = heritability of treatment-specific effect, $hqqe^2$ = heritability of treatment-specific epistasis effect.

RESULTS

QTS and QTT/P/M Mapping

There were 567 methylated genome loci used for mapping QTSs of two tobacco leaf traits (chromium content and total sugar content) with 60 phenotypes obtained from harvested leaves at three positions and different time points for three varieties grown in two locations. The QTS module in QTX Network was applied for mapping significant QTSs by setting two varieties and three locations as six treatments for detecting treatment-specific genetic effects.

The QTT/P/M module in QTX Network was applied to screen for significant RNA transcripts and to predict their genetic effects. A total of 2894 mRNA transcripts and 802 miRNA transcripts were used for QTT mapping. Similarly, QTP and QTM were applied to search significant proteins and metabolites. The concentration of 14 amino acids was measured for QTP mapping and 35 metabolites for QTM mapping accordingly. For QTS, QTP and QTM search, a total of 60 observations in six treatments were collected. The raw data of expression of RNAs, proteins and metabolites were transformed by standardization $(y - \mu) /$ for association mapping.

Association Mapping for Chromium Content Based on -Omics Data

There were a total of nine QTX loci (three for QTSs, four for QTTs, one for QTP, and one for QTM) that were detected as controlling chromium content in tobacco leaves (Table 1, Fig. 1 and Fig. 2). Three treatment-specific epistatic effects were identified between two QTSs with no individual effects. Large treatment-specific additive effects were found for QTSs, QTTs and QTP.

Table 1: Predicted QTX effects, QTX × environment interaction with significance and heritability for chromium content in tobacco leaf of three cultivars in two locations

Type	QTX[a]	Effect[b]	Predict	$-\log_{10}P$	h^2 (%)
QTS	Phm1376	q	0.43	10.05	20.29
		qe_1v_1	− 0.43	2.18	30.09
		qe_1v_2	− 0.37	1.72	
		qe_1v_3	− 0.41	2.00	
		qe_2v_1	0.45	2.38	
		qe_2v_2	0.60	3.78	
		qe_2v_3	0.78	6.09	
	Phm1053 × Phm1471	qqe_2v_1	0.41	2.09	11.24
		qqe_2v_2	0.47	2.55	
		qqe_2v_3	− 0.48	2.65	
QTT	mRNA2713	q	0.17	3.17	3.87
		qe_2v_1	0.37	5.11	24.62
		qe_2v_2	0.96	14.77	
	miRNA644	q	− 0.25	6.35	8.53
		qe_2v_1	− 0.47	5.79	16.29
		qe_2v_3	− 0.69	6.57	
	mRNA1119 × miRNA445	qq	0.43	17.75	26.25
QTP	Threonine	q	− 0.25	1.88	6.09
		qe_2v_3	− 0.63	2.69	6.21
QTM	Triacontanol	q	0.51	5.22	21.94

[a]mRNA2713: GenBank accession number FH405569; mRNA1119: GenBank accession number FH569168; miRNA644: sequence 5′-GCTTATCCATATTTGACCCGT TTTT-3′; miRNA445: sequence 5′-GAGCACGTACCCTGCTTCTCCA-3′.

[b]e_1: Guiding; e2: Xingyi; v_1: K326; v_2: Hongda; v_3: Zunyan 6.

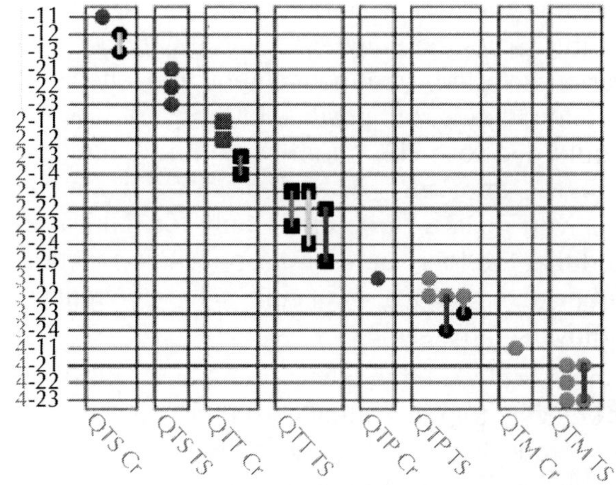

Figure 1: Network plot of detected QTXs for chromium content (Cr) and total sugar (TS) in tobacco leaves. Circle: QTX individual effect; Line between two QTXs: epistasis effects; Red: general effects for six environments; Green: environment-specific effects; Blue: general and environment-specific effects; Black: epistasis effects but without detected individual effects.

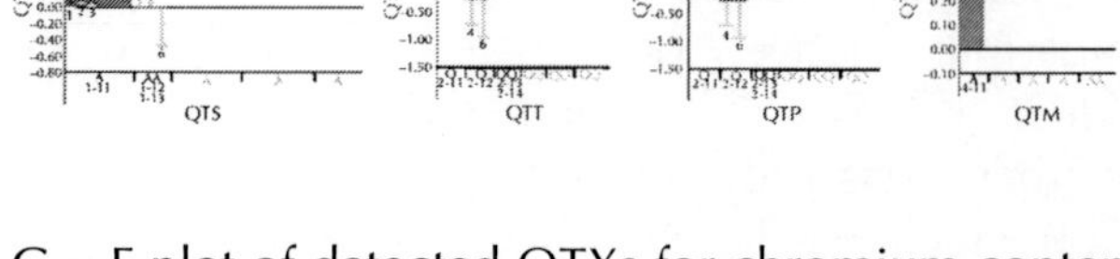

Figure 2: G × E plot of detected QTXs for chromium content (Cr) and total sugar (TS) in tobacco leaves. Red columns: general QTX effects for all six environments; Green lanes: environment-specific effects for $1 = e_1v_1$, $2 = e_1v_2$, $3 = e_1v_3$, $4 = e_2v_1$, $5 = e_2v_2$, $6 = e_2v_3$.

In the case of QTS, there were three methylated SNP loci (DArT markers) detected with significant additive (q), additive × treatment interaction (qe), and epistasis × treatment interaction effect (qqe) (Table 1). Phm1376 had a very significant additive effect ($-\log_{10}P$

= 10.05) and high heritability (hq^2 = 20.29%). The total additive × environment interaction had higher heritability ($hqqe^2$ = 30.09%). For the three varieties tested, treatment interaction effects of *Phm1376* were negative in Guiding, but positive in Xingyi. It was suggested that *Phm1376* could decrease chromium content in Guiding for all three varieties. *Phm1053* and*Phm1471* had epistasis × treatment interaction in the Xingyi with negative *qqe* effect only for cultivar Zunyan 6. It was indicated that the loci could have opposite impact on chromium content in tobacco leaves of a set of cultivars in different environments or various cultivars in the same environment.

For the analysis of QTT, there were two transcripts or unigenes and two miRNA detected for controlling chromium content of tobacco leaf (Table 1, Fig. 1 and Fig. 2). Unigene mRNA2713 had positive main effect (*q*) stable across environments and varieties. There was one miRNA (miRNA644) negatively associated with chromium content for two varieties in three environments. This QTT could also significantly decrease chromium content for K326 and Zunyan 6 in Xingyi. The data suggested that decreasing expression of mRNA2713 and increasing expression of miRNA644 could reduce chromium content in different varieties and environments. An epistasis between miQNA445 and mRNA1119 was detected for increasing chromium content stably across environments in three varieties. These results indicated that suppression of epistasis could significantly reduce chromium content in tobacco leaf.

In QTP mapping, only one amino acid (Threonine) was detected with significantly negative main effect (*q*) and treatment-specific effect (*qe*) in Xingyi for Zunyan 6 (Table 1, Fig. 1 and Fig. 2). It was suggested that threonine might actively participate in chromium degradation or be a byproduct of such activity when certain other factors presenting. Meanwhile, there was only one significant ($-\log_{10}P$ = 5.22) QTM (Triacontanol) which was identified with a positive main effect (*q*) on chromium content and high heritability (hq^2 = 21.94%) (Table 1, Fig. 1 and Fig. 2). This suggested that decreasing triacontanol content could reduce chromium content in tobacco.

Association mapping for total sugar content based on -omics data

Twelve loci were identified with the QTX module for association analysis between total sugar content (TS) and the four -omics datasets (three QTSs for methylated genes, five QTTs for mRNAs or miRNAs, four QTPs for amino acids, and three QTMs for metabolite elements (Table 2, Fig. 1 and Fig. 2)).

Table 2: Predicted QTX effects and QTX × environment interaction with significance and heritability for total sugar content in tobacco leaf of three cultivars in two locations

Type	QTX[a]	Effect[b]	Predict	$-\log_{10}P$	h^2 (%)
QTS	Phm1132	q	5.68	231.75	37.34
		qe_1v_1	− 5.79	42.16	16.27
		qe_1v_2	− 2.55	8.82	
		qe_1v_3	2.31	7.41	
		qe_2v_1	− 1.89	5.16	
		qe_2v_2	− 4.32	23.93	
		qe_2v_3	4.10	21.68	
	Phm1227	q	2.51	47.12	7.28
		qe_1v_1	− 2.85	10.89	21.07
		qe_1v_2	2.06	5.99	
		qe_1v_3	− 3.26	14.03	
		qe_2v_1	1.86	4.98	
		qe_2v_2	− 3.35	14.75	
		qe_2v_3	8.47	88.43	
	Phm1298	q	− 0.85	6.10	0.84
		qe_1v_3	− 7.59	71.61	14.31
		qe_2v_1	2.36	7.67	
		qe_2v_2	3.09	12.69	
		qe_2v_3	1.28	2.64	

QTT	mRNA537 × miRNA652	*qq*	− 0.63	9.90	22.87
	mRNA537 × miRNA775	qqe_1v_2	− 0.76	2.73	23.94
		qqe_1v_3	1.37	8.03	
	mRNA1218 × miRNA183	*qq*	− 0.42	4.97	10.44
		qqe_1v_1	0.89	2.68	18.44
		qqe_2v_3	− 1.05	3.27	
QTP	Threonine	*q*	0.97	1.63	0.85
	Lysine	*q*	− 7.24	62.55	46.9
	Lysine × serine	*qq*	− 2.19	7.40	4.31
	Lysine × phenylalanine	*qq*	6.10	53.47	33.27
		qqe_2v_2	− 2.16	4.21	0.7
QTM	Fructose	*q*	5.47	80.45	52.3
	Linolenic acid + linoleic acid	*q*	− 1.89	13.20	6.22
	Chlorogenic acid	*q*	1.15	5.32	2.31
	Fructose × linolenic acid + linoleic acid	*qq*	− 3.86	38.29	26.02
		qqe_1v_2	− 3.33	3.85	3.84
		qqe_2v_1	1.46	1.91	

Three methylated genome loci with significant additive (*q*) effects and additive × treatment interaction (*qe*) were detected (Table 2, Fig. 1 and Fig. 2). The locus Phm1132 had highly significant positive main effect ($-\log_{10}P = 231.75$) with high heritability ($hq^2 = 37.34\%$). For additive × treatment interaction, the Zunyan 6 had positive effects in the two locations, while the other two varieties had negative effects. It was suggested that *Phm1132* of Zunyan 6 could increase total sugar content across different environments. Though*Phm1227* was reported with significantly positive *a* effect ($-\log_{10}P = 47.12$), its heritability ($hq^2 = 7.28\%$) was not as large as *qe* interaction effects ($hqe^2 = 21.07\%$). Three cultivars tended to have opposite effects in the two locations. Zunyan 6 had quite large positive *qe* in the second location ($-\log_{10}P = 88.43$) while the other two cultivars did not. *Phm1298* had positive *qe* interaction for three cultivars in Xingyi, but significantly negative *qe* interaction for the Zunyan 6 variety in Guiding ($-\log_{10}P = 71.61$).

Three loci with epistasis between mRNA and miRNA genes also controlled total sugar content. mRNA537 had negative main epistasis ($hq^2 = 22.87\%$) jointly with miRNA652, and treatment-specific epistasis ($hqqe^2 = 23.94\%$) in Guiding (negative for Hongda and positive for Zunyan 6) together with miRNA775. mRNA1218 × miRNA183 had negative main epistasis ($hq^2 = 10.44\%$) and treatment-specific epistasis

($hqqe^2$ = 18.44%) in Xingyi for Zunyan 6. Therefore, epistasis might be useful as an efficient genetic tool for increasing total sugar content in tobacco leaf.

In QTP mapping, lysine was detected to have a large individual negative main effect (q) on total sugar content in tobacco leaves ($-\log_{10}P$ = 62.55 and hqq^2 46.90%), but positive epistasis effects (qq) along with phenylalanine ($-\log_{10}P$ = 53.47 and hqq^2 = 33.27%) (Table 2, Fig. 1 and Fig. 2). Meanwhile, for QTM mapping, fructose was detected with large positive individual effects (q) ($-\log_{10}P$ = 80.45 and $hqq^2$52.30%), while linolenic and linoleic acids had lower negative individual effects (q) ($-\log_{10}P$ = 13.20 andhqq^2 6.22%) (Table 2, Fig. 1 and Fig. 2). Epistasis effects of these two QTMs were also significant ($-\log_{10}P$ = 38.29 and hqq^2 26.02%).

DISCUSSION

QTX Network can Reveal Genetic Architecture of Complex Traits at Four -Omics Levels

The principal feat of this research was to implement QTXNetwork, a software program based on a mixed linear model, for analysis of -omics data. This research was able to take advantage of an abundance of data on gene methylation, transcript expression, protein content and metabolite characterization to find associations of QTS, QTT, QTP and QTM with two complex traits. Our goal in these analyses was to directly estimate the genetic effects of each type of loci on the genetic architecture of these traits. We believe this to be the first time that these new methods have been used to detect genome methylated loci, transcripts, proteins and metabolites associated with chromium content and total sugar content in tobacco leaves. The results showed that various types of genetic effects contributed to the two traits at different levels of -omics data, but that the composition and proportion of each type varied among -omics levels (Table 1 and Table 2). For example we observed that total heritability increased consecutively for genomic, transcriptomic, proteomic and metabolomics loci, which was consistent with the central genetic dogma of gene expression through transcripts and their resulting proteins and metabolites in the transfer of genetic information to phenotype.

Epistasis and Treatment Interaction may Play an Important Role in Complex Traits

Another discovery of this study was that the proportion of total heritability of epistasis and treatment interaction was very significant in the combination of trait and -omic evaluation, and that the total proportion of heritability based on epistasis and treatment interaction was nearly equal to that of the main factors. There was one QTS epistasis detected only in location 2 ($hqqe^2$ 11.24%) for chromium content among the four -omics levels. The proportions of total treatment interaction ($hqe_{+}qqe^2$) were 35.97%, 20.46%, 0.70% and 3.84% in genomic, trans criptomic, proteomic, and metabolic levels, respectively. In the case of total sugar content, the proportions of epistasis (hqq^2) were 37.58% and 26.02% for proteome and metabolism. However, no epistasis was detected for genome and transcriptome loci. In contrast, the proportions of epistasis and treatment interaction effects on heritability ($hqe_{+}qqe^2$) were 51.65%, but only 0.70% and 3.84% at the transcriptome, proteome and metabolome levels, respectively.

Application Potential of Detected QTXs in Crop Breeding

Molecular markers have enormous potential to improve the efficiency and precision of conventional plant breeding via marker-assisted selection [29]. The important challenge of applying genetic and -omics data to breeding is the identification of the genes underlying a trait of interest. We performed an integrated association mapping for chromium content and total sugar content based on genome, transcriptome, proteome and metabolites, and detected some QTSs, QTTs, QTPs and QTMs associated with two complex traits. The strategy to employ these molecular loci in the breeding practice should be considered prudently. For example, those QTX based on methylated loci of the genome were essentially directly useful as DNA markers and would be directly applicable in breeding practice.

In terms of marker assisted breeding for each of the two traits, chromium content could be selected with*Phm1376* which had a significant positive additive effect ($-\log_{10}P = 10.05$, and $hq^2 = 20.29$), indicating that demethylation of this locus could reduce chromium content for three varieties in two locations. The *qe*(additive by treatment interaction) effect of *Phm1376* was negative in Guiding for all three varieties tested in this study, but in Xingyi they were positive. This suggests that reduction of chromium content in tobacco leaves could be achieved by methylation of this locus in Guiding but demethylation of the same locus in Xingyi. The epistasis of Phm1053 and Phm1471 only had *qqe* effects in Xingyi, positive for K326 and Hongda, but negative for Zunyan 6, indicating that the best genotypic state in the two loci was demethylation for varieties K326 and Hongda, but methylation for Zunyan 6.

In the cases of the QTTs and QTPs associated with the traits of interest, there might be two strategies to use in practical plant breeding. For the first strategy, bioinformatics analysis can be carried out to make sure that the function of the transcript or proteins is based on a functional gene association with the investigated traits, and then the representative gene of the transcript can be developed as a DNA based marker useful for marker-assisted-selection. This strategy is more valuable when the transcripts or proteins have large genetic effects and high heritability. The second strategy would be based on direct use of the transcript as an indicative marker, where the abundance of the transcript would predict the performance of the genotype for the traits. In this study, two transcripts and two miRNAs presented association with chromium content. The transcripts representing mRNA2713 (GenBank accession numberFH405569) presented a stable positive main genetic effect (0.17) for the three varieties in two locations, with one unit reduction of transcript expression resulting in 0.17 units of chromium reduction. Another pair of transcripts, namely mRNA1119 (GenBank accession number FH569168) and miRNA445 (sequence: 5 -GAGCACGTACCCTGCTTCTCCA-3), presented a high and positive interaction effect (0.43) with moderate heritability (26.25%) and no environment interaction effect, indicating that they can be used as markers for breeding without concern for the specific growing environment. Another smRNA based locus, namely miRNA644 (sequence: 5 -GCTTATCCATATTTGACCCGTTTTT-3) showed a moderate negative main genetic effect (– 0.25) and heritability (8.53%), but presented a higher negative environment interaction effect (– 0.58 on average) and heritability (16.29%) in Xingyi, which indicated that this marker would be an environment-specific indicator of chromium content.

Finally, some metabolites had significant impacts on trait inheritance. For example, Lysine was detected with large individual negative main effects on total sugar content, but positive epistasis effects on this trait in combination with phenylalanine (Table 2). This indicated that high concentration of individual lysine could reduce the concentration of total sugar content in tobacco leaves. One explanation for this observation could be based on the Mallard reaction in vivo [30], which is the result of a chemical reaction between amino acids and reducing sugar. But when the two kinds of amino acids (lysine and phenylalanine) worked together as a pair, the joint effects (*qq*) were positive on total sugar content. Further study is required to confirm this and other associations of the two traits with the metabolomics, proteomic, transcriptomic and genome methylation datasets. Furthermore, the same kind of analysis could be used for additional traits that are of complex inheritance but for which biochemical (mRNA, protein and metabolite) analysis is very salient and as important as just a genome wide association test with random DNA markers.

ACKNOWLEDGMENTS

This study was partially supported by the National Basic Research Program of China (2011CB109306 and2009CB118404), the Program of Introducing Talents of Discipline to Universities of China ("111" Project, B06014), and Research Programs (CNTC-D2011100, CNTC-[2012]146, NY-[2011]3047, QKHRZ [2013] 02). We are grateful for editing from Dr. M. W. Blair and the suggestions of two anonymous reviewers.

REFERENCES

1. T.F. Mackay The genetic architecture of quantitative traits Annu. Rev. Genet., 35 (2001), pp. 303–339

2. L.X. Ku, W.M. Zhao, J. Zhang, L.C. Wu, C.L. Wang, P.A. Wang, W.Q. Zhang, Y.H. Chen Quantitative trait loci mapping of leaf angle and leaf orientation value in maize (*Zea mays* L.) Theor. Appl. Genet., 121 (2010), pp. 951–959
3. Z. Wang, X. Wu, Q. Ren, X. Chang, R. Li, R. Jing QTL mapping for developmental behavior of plant height in wheat (*Triticum aestivum* L.) Euphytica, 174 (2010), pp. 447–458
4. P. Langridge, D. Fleury Making the most of 'omics' for crop breeding Trends Biotechnol., 29 (2011), pp. 33–40
5. T.F. Mackay, E.A. Stone, J.F. Ayroles The genetics of quantitative traits: challenges and prospects Nat. Rev. Genet., 10 (2009), pp. 565–577
6. P. Martinez-Gomez, R. Sanchez-Perez, M. Rubio Clarifying omics concepts, challenges, and opportunities for *Prunus* breeding in the postgenomic era OMICS, 16 (2012), pp. 268–283
7. J.M. Thornsberry, M.M. Goodman, J. Doebley, S. Kresovich, D. Nielsen, E.T. Buckler *Dwarf8* polymorphisms associate with variation in flowering time Nat. Genet., 28 (2001), pp. 286–289
8. X. Huang, X. Wei, T. Sang, Q. Zhao, Q. Feng, Y. Zhao, C. Li, C. Zhu, T. Lu, Z. Zhang, M. Li, D. Fan, Y. Guo, A. Wang, L. Wang, L. Deng, W. Li, Y. Lu, Q. Weng, K. Liu, T. Huang, T. Zhou, Y. Jing, W. Li, Z. Lin, E.S. Buckler, Q. Qian, Q.F. Zhang, J. Li, B. Han Genome-wide association studies of 14 agronomic traits in rice landraces Nat. Genet., 42 (2010), pp. 961–967
9. J.A. Rafalski Association genetics in crop improvement Curr. Opin. Plant Biol., 13 (2010), pp. 174–180
10. F. Tian, P.J. Bradbury, P.J. Brown, H. Hung, Q. Sun, S. Flint-Garcia, T.R. Rocheford, M.D. McMullen, J.B. Holland, E.S. Buckler Genome-wide association study of leaf architecture in the maize nested association mapping population Nat. Genet., 43 (2011), pp. 159–162
11. O. Zuk, E. Hechter, S.R. Sunyaev, E.S. Lander The mystery of missing heritability: genetic interactions create phantom heritability Proc. Natl. Acad. Sci. U. S. A., 109 (2012), pp. 1193–1198
12. L.S. Yung, C. Yang, X. Wan, W.C. Yu GBOOST: a GPU-based tool for detecting gene–gene interactions in genome-wide case control studies Bioinformatics, 27 (2011), pp. 1309–1310

13. G.K. Chen, K. Wang, A.H. Stram, E.M. Sobel, K. Lange Mendel-GPU: haplotyping and genotype imputation on Graphics Processing Units Bioinformatics, 28 (2012), pp. 2979–2980

14. P. Winterhalter, R.L. Rouseff Carotenoid-derived Aroma Compounds An American Chemical Society Publication (2002)

15. Bombarely, K.D. Edwards, J. Sanchez-Tamburrino, L.A. Mueller Deciphering the complex leaf transcriptome of the allotetraploid species *Nicotiana tabacum*: a phylogenomic perspective BMC Genomics, 13 (2012), p. 406

16. Lei, X.H. Zhao, K. Zhang, J. Zhang, W. Ren, Z. Ren, Y. Chen, H.N. Zhao, W.J. Pan, W. Chen, H.X. Li, W.Y. Deng, F.Z. Ding, K. Lum Comparative transcriptome analysis of tobacco (*Nicotiana tabacum*) leaves to identify aroma compound-related genes expressed in different cultivated regions Mol. Biol. Rep., 40 (2013), pp. 345–357

17. B.G. Xiao, X.P. Lu, F.C. Jiao, Y.P. Li, Y.H. Sun, Z.K. Guo Preliminary QTL analysis of several chemical components in flue-cured tobacco (*Nicotiana tabacum* L.) Acta Agron. Sin., 34 (2008), pp. 341–349

18. C. Goulet, C. Goulet, M.C. Goulet, D. Michaud 2-DE proteome maps for the leaf apoplast of *Nicotiana benthamiana* Proteomics, 10 (2010), pp. 2536–2544

19. E. Julio, B. Denoyes-Rothan, J.L. Verrier, F.D. de Borne Detection of QTLs linked to leaf and smoke properties in *Nicotiana tabacum* based on a study of 114 recombinant inbred lines Mol. Breed., 1 (2006), pp. 69–91

20. G. Bindler, J. Plieske, N. Bakaher, I. Gunduz, N. Ivanov, R. Van der Hoeven, M. Ganal, P. Donini A high density genetic map of tobacco (*Nicotiana tabacum* L.) obtained from large scale microsatellite marker development Theor. Appl. Genet., 123 (2011), pp. 219–230

21. H. Cui, S.T. Zhang, H.J. Yang, H. Ji, X.J. Wang Gene expression profile analysis of tobacco leaf trichomes BMC Plant Biol., 11 (2011), p. 76

22. W.Q. Lui, H.X. Guo, H. Li Proteomics identification of differentially expressed proteins relevant for nicotine synthesis in flue-cured tobacco roots before and after decapitation Agric. Sci. China, 9 (2008), pp. 1084–1090

23. X.P. Lu, Y.J. Gui, B.G. Xiao, Y.P. Li, Z.J. Tong, Y. Liu, X.F. Bai, W.R. Wu, L. Xia, E. Huttner, A. Kilian, L.J. Fan Development of DArT markers for a linkage map of flue-cured tobacco Chin. Sci. Bull., 58 (2012), pp. 641–648

24. Irizarry, M. As trend, T.P. Speed A comparison of normalization methods for high density oligonucleotide array data based on variance and bias Bioinformatics, 19 (2003), pp. 185–193

25. R.A. Irizarry, B. Hobbs, F. Collin, Y.D. Beazer-Barclay, K.J. Anton Ellis, U. Scherf, T.P. Speed Exploration, normalization, and summaries of high density oligonucleotide array probe level data Biostatistics, 4 (2003), pp. 249–264

26. Y. Guo, H. Liu, Z. Yang, J. Chen, Y. Sun, X. Ren Identification and characterization of miRNA me in tobacco (*Nicotiana tabacum*) by deep sequencing combined with microarray Gene, 501 (2012), pp. 24–32

27. M.B. Eisen, P.T. Spellman, P.O. Brown, D. Botstein Cluster analysis and display of genome-wide expression patterns Proc. Natl. Acad. Sci. U. S. A., 95 (1998), pp. 14863–14868

28. J. Yang, J. Zhu, R.W. Williams Mapping the genetic architecture of complex traits in experimental populations Bioinformatics, 23 (2007), pp. 1527–1536

29. B.C. Collard, D.J. Mackill Marker-assisted selection: an approach for precision plant breeding in the twenty-first century Philos. Trans. R. Soc. Lond. B Biol. Sci., 363 (2008), pp. 557–572

30. H. Nersten The Maillard Reaction: Chemistry, Biochemistry and Implications The University of Reading, Reading, UK (2005)

Chapter 2

Prognosis for Genetic Improvement of Yield Potential and Water-Limited Yield of Major Grain Crops

Antonio J. Hall[a] and Richard A. Richards[b]

[a]IFEVA, Facultad de Agronomia-Universidad de Buenos Aires/CONICET, Av. San Martín 4453, C1417DSE Buenos Aires, Argentina
[b]CSIRO Plant Industry, PO Box 1600, Canberra, ACT 2601, Australia

ABSTRACT

Genetic improvement of crop yields under potential (Yp) and water scarce conditions (Yw) will be an important avenue to improved food security over the next four decades, at the end of which projected demand for food, feed and biofuel feedstock is expected to level out. Current measured *relative* rates of improvement in Yp and Yw for the three main cereal crops (maize, wheat and rice) in many cropping systems in the world are mostly well below the 1.16–1.31% y^{-1} rates required to meet projected demand for cereals in 2050. These relative rates can be expected to fall further if the current *absolute* rates of yield improvement continue unchanged and/or the current indications of stagnation in yield improvement for some crops in some regions of the world become widespread. This review assesses the available evidence for unexploited opportunities for enhancing current rates of genetic improvement for Yp and Yw, and examines some substantive proposals for achieving the same end through genetic engineering of photosynthesis, above-ground ideotype design, and improvement of root capacity for water uptake. Because time is of the essence, special attention is paid to the time scales required to progress potentially useful traits through to proof of concept under field conditions, from there to farmer-ready cultivars, and for widespread adoption by farmers of the improved cultivars. The requirements of breeders for inclusion of potentially important traits into the conventional breeding process are outlined and the value of molecular breeding tools as aids to genetic improvement of simple and complex traits is considered. Intellectual property and regulatory requirements are taken as additional potential drags on the rates of adoption into the breeding process of useful traits and/or to the free flow of information between research teams. Main conclusions are: (a) there are some unexploited opportunities to improve Yp or Yw in some crops and/or cropping systems; (b) exploitation of these opportunities could be hastened by increasing funding for focussed research and by identifying and eliminating or reducing drags at various stages of the idea to farmer-ready cultivar chain; (c) the time-scales required for major improvements in yield in farmer-ready cultivars through genetic engineering or ideotype realisation are likely to be measured in decades rather than years; (d) current and expected future relative rates of progress in Yp and Yw are a matter of real concern and are insufficient to meet projected demand for cereals by 2050. Possible step changes in Yp or Yw powered by genetic improvements such as exploitation of hybrid vigour in rice or hypothetical successes in the genetic engineering of photosynthesis are unlikely to change this outlook. The pessimistic assessments embodied in the last two conclusions should not obscure the fact that without continued investment in the search for sources of genetic improvement and development of aids to breeding, the unmet demands for cereals in 2050 will be even greater.

INTRODUCTION

Context

Estimates of world population for 2050 are in the order of nine billion people (Lutz and KC, 2010) resulting in an increased demand for food and feed. Additional weighty drivers of demand for food and feed grains over the next four decades include improved standards of living in many developing nations (Koning et al., 2008 and Cirera and Masset, 2010), policy initiatives aimed at reducing current levels of malnourishment (Bruinsma, 2009), and government policies that encourage the use of biofuels derived from grains (Fischer, 2009a). Estimating the impacts of these drivers on demand for grain and food by 2050 is a complex task hemmed in by uncertainties, but detailed analyses suggest that an increase (with respect to 2006 values) of the order of 49% for cereals (Bruinsma, 2009) will be required to meet expected demand. Estimates of use of cereals as feedstock for first generation biofuels are even harder to formulate, butFischer (2009a) has estimated (for scenarios that appear reasonably likely) that an extra amount (over and above that required for food and feed) of between 163 Mt and 363 Mt y^{-1} will be needed for 2050, equivalent to an additional increase in global demand for cereal grains of between 9% and 19% at that time. Meeting these demands for cereals would require global harvests to increase at annual compound rates of 1.16% y^{-1} (lower biofuel requirement estimate) or 1.31% y^{-1} (upper biofuel requirement estimate) from now until 2050.

There is a developing, albeit broad, consensus within the scientific and policy-making communities that only a small proportion of the required increase in food production can come from expansion of the area that is currently cropped (Koning et al., 2008 and Bruinsma, 2009). Limited availability of currently unexploited agriculturally suitable soils (Smith et al., 2010 and Fischer, 2009a), a growing appreciation of the magnitude of the release of CO_2 from soil organic matter that inevitably attends the conversion of grassland or forest to cropping (Spink et al., 2009 and Searchinger et al., 2008), a strong desire to preserve hitherto untouched portions of important ecosystems such as tropical rainforest, and a realisation of the dangers inherent in expanding agriculture to fragile ecosystems (Cassman et al., 2003) are important elements in this developing consensus. Additionally, it is expected that there will be losses of good quality agricultural soil for urban development and other non-agricultural uses (Satterthwaite et al., 2010 and Smith et al., 2010). Finally, there is a growing realisation that opportunities for expanding irrigation (a key factor in the increases in food production over the last 50 years) are limited and that in at least some production systems water currently used for irrigation will have to be reduced to favour other uses such as industrial or urban consumption (Strzepek and Boehlert, 2010). A hard to quantify factor which may impact on food production in the immediate future is the reduction in inputs used by farmers in developed countries, either because of increased relative prices for fertilisers or because of regulatory restrictions aimed at reducing losses of reactive nitrogen from fields into the broader environment. Berry and Spink (2006) have suggested that changes of this nature may be operating in determining an increasing yield gap between farmer and potential yields for oil seed rape in the UK, and Brisson et al. (2010) include this factor as a possible explanation of yield stagnation for wheat in France.

The above analyses lead to the conclusion that most of the increased food production over the next four decades will have to be achieved through increased crop yields on currently exploited agricultural soils. In some developing countries, especially in Africa, there is considerable scope for improving yields through application of well-understood technologies such as fertilisation (Cooper et al., 2008 and Tittonell et al., 2008) but effective implementation of this strategy will also require considerable off-farm changes such as infrastructure improvement, increased availability of inputs for agriculture and effective markets for agricultural produce, as well as diffusion of know how and a reduction of risk aversion among farmers. Successful realisation of the "Africa option", of itself, would not suffice to cover the required increase to 2050 and would, necessarily, take decades to complete.

There has been some considerable debate in the literature as to whether current rates of yield improvement will suffice to meet the increased demand for 2050. This debate has several dimensions, including spatial measurement scale (i.e., global vs. country vs. region), crop species (e.g., wheat vs. rice vs. maize), and benchmark used (i.e., farmer yields derived from national or regional statistics, historical yield trial data, side-by-side comparisons of cultivars released at different times in the past, and crop simulation modelling) and conclusions reached are strongly dependent on scale, crop species, approach used and country involved. Further important dimensions of this debate are whether yield progress shows signs of plateauing and whether the relative rates of increase observed in crops and regions in which yields are increasing are sufficient to meet the 1.16–1.31% y^{-1} compound rates needed to satisfy projected demand in cereals for food, feed and biofuels for 2050 (Bruinsma, 2009 and Fischer, 2009a). In wheat (Calderini and Slafer, 1998, Acreche et al., 2008, Brisson et al., 2010, Graybosch and Peterson, 2010,Cassman et al., 2010 and Mackay et al., 2011) and rice (Peng et al., 1999) or in both crops (Cassman et al., 2010 and Fischer and Edmeades, 2010), there is strong evidence that, at least in some countries (regions) of the globe, yield is plateauing and/or relative rates of yield progress are falling below 1.16% and can be expected to fall further if current trends persist. For maize, and particularly in the US, the situation is somewhat less clear. Some analyses (e.g., Phillips, 2010, country wide data; Fischer and Edmeades, 2010, Iowa) suggest continued increases in farmer yield. On the other hand, Grassini et al. (2011), who focus on irrigated high yield environments in Nebraska, present data that indicate farm yields are approaching or have reached their economic upper limit, quite close to simulated potential yield values. It also needs to be noted that Fischer and Edmeades (2010) estimated a relative rate of progress in potential yield for maize in Iowa of 1% y^{-1} (just outside the range of 0–0.9% y^{-1} they estimate for wheat and rice), and Cassman et al. (2010) suggest a yield plateau may be developing for irrigated maize in the US. The hard truth emerging from these different approaches, scales and crops is that the vast majority of data indicate relative rates of potential yield progress fall below the necessary exponential rate required to meet projected demand for 2050 and, in a worrying number of countries and regions, that there is strong evidence of yield plateaus. Some authors (e.g., Jaggard et al., 2010 and Ewert et al., 2005) have assumed that current *relative* rates of yield progress, as derived from linear functions fitted to historical yield data and estimated for the end point of their data series, will be maintained in the future. This seems unlikely, even in the absence of insuperable biophysical limitations. A continuation of current linear trends in yield progress are, almost certainly, the best possible outcome, and this implies a continuously decreasing, rather than a constant relative rate of yield progress.

The routes to the achievement of increased and sustainable global food security by 2050 are several (e.g.,Godfray et al., 2010a and Godfray et al., 2010b suggest five) and include genetic improvement for grain crop yield potential in unstressed environments (Yp) and for water-limited yield potential in dryland (rainfed) environments (Yw). In this paper we attempt to assess the current status, opportunities and requirements for progress and its verification in both Yp and Yw, paying particular attention to likely timescales. We concentrate on the most important cereals (maize, rice, wheat) because of their dominant contribution to current and likely future global food, feed and biofuel requirements.

Genetic Improvement of Yield

Breeding is at the heart of any process aimed at improving Yp or Yw, and is currently undergoing a rapid expansion of the technical options available to progress any given set of aims, including Yp and Yw. Consequently, we give due consideration to the nature and practical requirements of breeding processes aimed at achieving farmer-ready cultivars with improved Yp or Yw. Genetic improvement may be further enhanced by new or improved traits made available through genetic engineering or the mining of existing genetic variability. The identification of these traits and the importance of them should be assessed by whether they fit into frameworks that are appropriate to the improvement of either Yp or Yw, so we make some attempt to deal with this issue. With the 2050 deadline and germplasm development and breeding timescales very much in mind, we have used some published estimates of the likely time required to improve, via genetic engineering, the photosynthetic capacity of crops to illustrate this aspect of the road ahead. In a similar vein, we have documented the time required, in the past, to progress some ideas or discoveries of traits related to improved Yw up to the proof of concept or cultivar release stages. Time is also required for improved cultivars to be widely adopted by farmers, and we briefly touch upon this issue.

The range of traits that have been suggested as possible contributors to the improvement of either or both of Yp and Yw is very large, and we make no attempt at comprehensive coverage of these suggestions. Instead, we have used, as examples of the difficulties that must be faced, two particular processes which we believe are potentially very important and, at the same time, constitute very large challenges for progress to improved yields. Firstly, improvement of photosynthetic capacity for both non-stressed and water-limited conditions, and secondly, improvement of root system capacity to capture water in dry environments.

We recognise that the degree of success that might be achieved in Yp and Yw will be affected by climate change, especially via changes in atmospheric CO_2 and O_3 levels, mean temperatures and frequencies of episodes of heat and water stress (Gornall et al., 2010). These features of climate change will impact on issues such as sowing dates and crops best suited to particular environments; and current rates of progress (or lack thereof) estimated for C3 cereal crop yields across the world include some degree of response to changes in temperature and CO_2. The knowledge base underpinning attempts to predict possible effects of changes in atmospheric CO_2 and O_3 levels on crop productivity (e.g., Fischer, 2009a and Jaggard et al., 2010) is limited (Gornall et al., 2010), as is the case for acute or chronic changes in temperature (cf. Hall and Sadras, 2009). Breeders and farmers have, in the past, been successful in optimising their activities over broad spatially determined ranges in environmental conditions (Cooper et al., 2008). Effects of climate change will develop slowly over time, so business-as-usual genetic improvement and gradual adaptation of farming systems are likely to handle these effects successfully. In this review we do not deal with the possible effects of climate change on genetic improvement for Yp and Yw.

GENETIC IMPROVEMENT OF YIELD POTENTIAL

Rates of Yp Increase: Current Status

Any assessment of the prospects for genetic improvement needs to consider what are the current rates of increase in Yp. In what follows, we consider this issue in the context of some of the available empirical (i.e., not theoretical) evidence pertaining to crop species as we know them today, that is, without the possible benefits of bio-engineered photosynthesis. Rather than a comprehensive review, we have concentrated on the main cereals but have, on occasion, had recourse to other crops. Available empirical evidence paints different pictures for each of the main cereals and highlights the fact that the quality of the evidence varies between cereals.

Duvick and Cassman (1999) examined evidence from various sources (including side-by-side yield comparisons of maize hybrids released at different times in rainfed trials, trends in average yields for Iowa and Nebraska, irrigated yield contest results in the same states), and came to the conclusion that while there was evidence for continued increases in Yw, there was no substantive indication of improvement in Yp of maize hybrids used in the region in the 15 years preceding 1999. Cassman et al. (2010) found an incipient plateau in irrigated maize yields for the western Corn Belt and Great Plains states for the 2002–2008 period, and Grassini et al. (2011) report that model simulations for mean yields of current maize hybrids under optimum water and nitrogen availability in Nebraska are of the order of 17.5 t ha^{-1} over the 1988–2008 period, not very different to the mean of 18.2 t ha^{-1} reported by Duvick and Cassman (1999) for irrigated contest-winning yields in the same state for 1983–1997. Messina et al. (2009) provide data that suggest that under full irrigation the yield of maize in a target population of environments was increasing at 0.79% y^{-1} in 2007, although other less optimistic interpretations of their data are also possible. Taken together, this evidence suggests that Yp of maize under intensive irrigated farming in the US has either not changed over the last two decades or is increasing at a rate insufficient to meet expected demand.

Estimates of yield potential for maize in other regions of the world are hard to come by (cf. Fischer and Edmeades, 2010). However, Luque (2000) compared yields of 7 maize hybrids, released between 1965 and 1997 in Argentina, in side-by-side experiments under irrigation and a range of crop population densities over 2 years. Average yields at crop population densities of 9 pl m^{-2} (none of these hybrids exhibited negative yield responses at this population density) increased steadily with year of release from 7.4 to 14.8 t ha^{-1}. These increases were associated with longer times to anthesis, greater stay green, greater anthesis RUE, improved partitioning to the ear around flowering, shortened anthesis-silking interval, and better grain set. It may also be important that over the period in which these hybrids were released (almost exclusively for use in rain-fed agriculture) there was a continued process of introgression of US dent germplasm into the flint germplasm that used to characterise Argentine maize and a shift from double hybrids to single hybrids (Luque et al., 2006). Although this result from Argentina is a single example it suggests that, at least in some maize-producing regions of the world, there may be some remaining scope for improving maize Yp.

The standard way of determining relative rates of genetic improvement is to grow historic sets of varieties side by side and to control pests and diseases so as to assess true genetic differences in grain yield, although recent statistical advances in the analysis of unbalanced data sets has also been very important (Smith et al., 2005 and de la Vega and Chapman, 2010) and may provide an alternative route that avoids the effects of loss of yield potential of cultivars with time from year of release, as has occurred in rice (see below) and sunflower (de la Vega et al., 2007). Yield potential in irrigated spring wheat, as determined in experiments in the Yaqui Valley in Mexico, was increasing at 0.3% y^{-1} in 2005 (Fischer and Edmeades, 2010 and Sayre et al., 1997). However, in the warmer environment of Tlatizapán, yields were less than half those of the Yaqui Valley and showed little tendency to increase in cultivars released after 1966 (Reynolds et al., 2000). At both sites there was evidence that higher yields were associated with higher stomatal conductance, lower canopy temperature and greater photosynthesis (Fischer et al., 1998 and Reynolds et al., 2000). Historic set studies on irrigated winter wheats from different periods in the main winter wheat regions in China (Zhou et al., 2007 and Xiao et al., 2012) have demonstrated mean relative rates of yield increase of 0.68% y^{-1} and 0.75% y^{-1}, estimated at the year of release of the most recent cultivar.

The most comprehensive data sets for wheat are from the UK where environments are very favourable for temperate species as there is generally adequate rainfall and the season is mild, so that results from rainfed trials may serve, in most years, as an approximation to true yield potential. In the UK national trials are conducted annually contrasting the latest cultivars. These have complete pest and disease protection and agronomic advances can be separated from improvements due to genetics. Data from rainfed variety trials in the UK and France show consistent upward trends in yields (Spink et al., 2009, Jaggard et al., 2010 and Brisson et al., 2010). Mackay et al. (2011) conducted a comprehensive analysis of genetic gain in winter wheat in the UK since 1947 which involved 3590 site-year combinations. They found a linear rate of yield increase of 69 kg $ha^{-1}y^{-1}$ between 1948 and 2007, a value that translates into a relative rate of 0.76% y^{-1} in 2007. They also found that since 1982 at least 88% of improvement in yields across a range of species could be attributed to genetic improvement. Studies on varieties released in the UK after 1990, which led to an estimated relative rate of genetic gain of 0.59% y^{-1} (Fischer and Edmeades, 2010), showed that the increase in yield has been associated with greater above-ground biomass accumulation, greater pre-anthesis RUE and higher stem water-soluble carbohydrate content at anthesis (Shearman et al., 2005). In spite of the fact that water stress is neither frequent nor intense in the UK, some caution needs to be exercised in assigning the observed yield trends in rain-fed yield trials exclusively to increased yield potential, leaving aside an improved stress tolerance component, as suggested for maize in the US (Duvick and Cassman, 1999) and by Sadras and Lawson (2011) for rainfed wheat in the harsher environment of South Australia. Experiments under irrigation would be helpful in clarifying this uncertainty.

Improvements in yield must come from either increases in crop biomass or harvest index (HI) or from a combination of both of these. Many studies on historic wheat sets up to the 1980s failed to show any increase in crop biomass and the main increases were attributable to an increase in HI (often as a result of earlier flowering). Furthermore, grain number per unit area is the yield component that has been responsible for this increase and there has been little change in kernel weight. However, in the most recent studies on historic data sets in the UK, China, Mexico and Australia (referenced above) there is increasing evidence that increases in crop biomass are beginning to contribute to increased wheat yields.

A careful study aimed at establishing genetic gain in inbred *indica* rice cultivars in the Philippines (Peng et al., 1999, Peng et al., 2000 and Peng et al., 2010), using side-by-side trials of cultivars released at different times after 1968, indicates that there has been no progress in potential yield. Rather, breeders efforts have served to maintain a yield potential equivalent to that of the earliest cultivar (IR8, released in 1968, and which currently yields a good deal less than at the time of its release) in the face of some evolving (and, as yet, unidentified) biotic or abiotic challenges. Ideotype-based approaches, backed by evaluation of reputed traits of interest using simulation modelling (e.g., Aggarwal et al., 1997), have been used in attempts to develop new plant type (NPT) inbred cultivars, but there appears to be little, or no, realised gain in potential yield in these NPTs with respect to earlier inbred cultivars (e.g., Yang et al., 2007). Horie et al. (2006)demonstrated that grain yields across a set of nine inbred cultivars over 2 years were associated with canopy conductance, a finding which may point to the existence of some exploitable genetic variability for yield, but this finding has not, to our knowledge, been tested in a breeding program nor linked to genetic gain in rice Yp.

The yield advantage of hybrid rice is typically 10–15% more than inbred cultivars (e.g., Peng et al., 1999,Yang et al., 2007, Bueno and Lafarge, 2009, Lafarge and Bueno, 2009, Bueno et al., 2010 and Li et al., 2009a). This advantage has been enough to overcome the economic limitations imposed by the greater (with respect to inbreds) annual replacement cost of seed. Hybrid rice now covers close to 60% (mean for 2004–2008) of the area sown to rice in China (Li et al., 2009a), and the area is also expanding, albeit at a slower rate and from a much lower base, in south-east and south Asia (Janaiah and Xie, 2010). Li et al. (2009a) provide data which show that commercial hybrid rice yields have been growing at a steady 46.4 kg ha^{-1} y^{-1} over the last 25 years (i.e., relative rate of yield advance of 0.64% y^{-1} for 2008). In a comparison involving sets of four hybrids and four inbreds over two seasons, Bueno and Lafarge (2009)found mean hybrid biomass at harvest and mean hybrid HI to be greater than the corresponding values for inbred lines.

Annual replacement cost of seed is especially important in crops such as rice and wheat, which are sown at a much higher crop population density than species such as maize and sunflower in which the introduction of hybrids has been an effective route to increased yield. In contrast to rice there is negligible hybrid wheat grown commercially despite the large private investment in global breeding. In a comparison of hybrid and pureline entries in performance nurseries conducted over 20 years in the southern Great Plains of the US the average yield advantage of the wheat hybrids was about 5% (Koemel et al., 2004). It is likely that the seed production and breeding costs are still not competitive with pureline varieties in wheat.

In summary, empirical evidence for inbred tropical rice and hybrid maize in environments in which intensive agriculture is practiced and important breeding efforts have been made, strongly suggests that there has been little or no change in Yp over the last few decades. There may be opportunities to increase maize Yp in some regions of the world thanks to introgression of previously unused germplasm, but further research on this issue is required. Genetic improvement of yield potential of inbred tropical rice seems to have plateaued, although hybrids may offer a one-off step-increase. Finally, the difficulties in realising the expected yield advantages of NPT rice may constitute an illustration of the difficulties that must be overcome in ideotype-based approaches to yield improvement. In contrast to maize and rice, there is a fair amount of evidence consistent with a continuing increase in Yp in wheat. However, current rates of progress in this crop, measured in a range of environments across the world, are consistently lower than that of the 1.16–1.31% y^{-1} that would be required to meet projected demands for 2050.

Suggested Options for Future Increases in Yp

Suggested responses to the challenge of increasing yield to meet the 2050 deadline vary considerably in focus and in degree of detail. Some are very generic and, very importantly, often lack time scales for implementation (e.g., Phillips, 2010, Zhang, 2007 and Dunwell, 2011) and receive no further attention here.

Other proposals to increase Yp, which might be termed top-down or from crop-to-attribute (e.g., Sylvester-Bradley et al., 2005, Berry and Spink, 2006 and Spink et al., 2009), are based on current understanding of the ecophysiology of crop radiation capture, the efficiency with which captured radiation is transformed into biomass and the proportion of that biomass converted into grain (i.e., the Monteith (1977) framework), and attempt to design ideotypes that would be able to capture more resources and channel them to yield. Others (bottom-up, or process to crop) focus more narrowly on bio-engineering the photosynthetic process (Murchie et al., 2008, Zhu et al., 2010, Hibberd et al., 2008 and Furbank et al., 2009). Still others have a broader remit and seek to cover the full spectrum of the bottom-up and top-down approaches, expanding these to include the use of genetic resources from related and unrelated species (Reynolds et al., 2011,Parry et al., 2011, Foulkes et al., 2011 and Spink et al., 2009). Taken together, these proposed frameworks lay out the complexity of the challenges that need to be overcome in order to accumulate the desired traits in elite breeding material that is reasonably free of linkage drag and which might be used to diffuse these traits into locally adapted cultivars fairly quickly. Importantly, several of these contributions attempt to establish possible time-frames for achieving their various targets (e.g., Zhu et al., 2010, Spink et al., 2009 and Reynolds et al., 2011). Spink et al. (2009) suggest 5- to 15-year periods might be needed to improve such traits as delayed canopy senescence, early canopy closure and earlier stem extension in wheat; and bioengineering of photosynthesis might take between 5 and 30 years (see Section 2.3).

We believe that some degree of caution is warranted with respect to the possible time frames referenced above. Reasons for caution include limited success and slow rates of progress in past attempts to improve crops for some physiological traits (e.g., Richards, 2006). Slow rates are mainly attributed to the time required to develop fast and effective selection methods, develop appropriate breeding germplasm and then produce sufficient seed of contrasting lines to obtain proof of concept in the field. Further development will then be dependent on plant breeders having the resources, time and commitment to integrate the physiological trait into genetic backgrounds that are likely to be adopted by industry. During this time breeders will need to deal with possible trade-offs between traits and matching trait to specific environments (i.e., managing genotype by environment interactions). All of these reasons point to longer terms for the achievement of some useful improvement in farmer-ready cultivars with greater Yp.

Photosynthesis: the Big Bazooka in the Yp Game?

Here we consider the prospects and likely time scales for improving yield potential via bioengineering of photosynthesis in C3 crops and/or exploitation of currently available sources of genetic variation for photosynthetic traits. Photosynthesis is the obvious process to consider given that any improvement in it could translate to other crops and to most if not all environments. Improvements in photosynthesis have the potential to translate to higher yields under favourable conditions through improved RUE and also under water-limited conditions through improved water-use efficiency (WUE). Despite breeding efforts world-wide to improve grain yield of all our crops there is only limited evidence to suggest that breeding has increased photosynthesis. Most increases in yield of our main crops achieved by breeding since around 1900 have been due to increases in HI (Fischer and Edmeades, 2010). Maize may be an exception here as the increase in maize yield in southern Canada between 1959 and 1988 was due to greater biomass and an extended period of photosynthesis after flowering (Tollenaar, 1991). Although clearly photosynthesis has increased over time, mainly as a result of improved or extended seasonal radiation capture; increased amounts of nitrogen fertiliser applied to crops has greatly contributed to this as well.

Other factors contributing to the importance of improving photosynthesis include: (a) the likelihood that fertiliser use in high production systems has already been optimised, and (b) that CO_2 enrichment experiments in the field demonstrate that increasing photosynthesis translates to higher grain yields (Long et al., 2006). Several recent papers (Parry et al., 2011, Zhu et al., 2010 and Murchie et al., 2008) have comprehensively reviewed the prospects of improving photosynthesis or have focused on the subplot of conversion of the photosynthetic carbon fixation process from its C_3 to its C_4 variant (Hibberd et al., 2008 and Furbank et al., 2009). Collectively, they lay out a very broad range of approaches of varying complexity aimed at increasing crop biomass production (and hence yield) that span the mining of existing germplasm for traits such as staygreen and spike photosynthesis through to very complex bioengineering of both the biochemical and photochemical phases of photosynthesis and of its associated processes (e.g., photorespiration).

Transgenic approaches aimed at C_3 to C_4 conversion at biochemical and anatomical levels and changes in Rubisco properties are obviously the most difficult to achieve. Here, we do not go over this ground again, but focus on the complexities associated with the various approaches, as perceived by these authors, and the timescales which have been suggested for achieving the various goals (e.g., Table 1 in Murchie et al., 2008, Table 2 in Zhu et al., 2010; and Table 2 in Parry et al., 2011). Perhaps the first thing that needs to be said is that the attempts made in these papers to categorise the proposed approaches according to their degree of complexity and possible impact and, in two of them, to suggest time scales for realisation of the various goals is a very welcome change from generic calls to action to bioengineer photosynthesis or naive claims of the potential importance for global food security of the elucidation of one piece of the photosynthetic process jigsaw puzzle.

In the context of this review, it is important to highlight the fact that estimates of the predicted impact of the various proposed changes in the photosynthetic processes on crop carbon assimilation are mostly in the 5–30% range and time required to achieve success in these assorted objectives from 5 to 20 years. Not surprisingly, there are some differences in forecasts between authors. Thus Zhu et al. (2010) predicts the impact of Rubisco with low or nil oxygenase activity and the C_3 to C_4 conversion in the order of 30% whileParry et al. (2011) suggest 100% for the first and 50% for the second, and increased mesophyll conductance is suggested to require 20 y with high investment by Zhu et al. (2010) but only 5 y by Parry et al. (2011). Zhu et al. (2010) caution that 10–20 additional years may be needed to incorporate any successful changes into farmer-ready cultivars; but Parry et al. (2011) suggest that 5 y might be needed to mine existing germplasm for useful traits and introgress them into wheat lines available for first crosses in breeding programs. These timescales would seem very optimistic for food crops even for relatively simple traits (see, this section and Section 3.4).

The whole issue of strategies for improving photosynthetic performance of crops through bioengineering of the basic processes of carbon fixation is, of necessity, rather speculative. However, the complexity of the challenge is clearly very great and the time needed to achieve progress (in the form of farmer-ready cultivars) will likely be in the order of 20–30 y for changes that could signify improvements in photosynthetic performance of C3 crops of 15–30%, always providing that realisation of these bioengineering goals does not bring into play unconsidered trade-offs that often complicate the transition in scales from process to crop (Sinclair et al., 2004 and Passioura, 2010).

A further, often unconsidered, dimension of the challenge to improving grain yields through improved crop photosynthetic performance is the time required for new cultivars to be widely adopted. These intervals can vary considerably with crop and context. Glyphosate-resistant soybean covered practically the whole of the production area of this crop in Argentina in 5 years (Vara, 2004);

reaching the current ca. 60% coverage of the rice production area in China with hybrid rice took 15 years (Li et al., 2009a); and spread of GM corn in the US from 1994 onwards took place at a linear rate of 5.8% of sown area per year (Phillips, 2010).

To highlight the magnitude of the challenges that need to be overcome in order to raise Yp to the levels required to meet demand expected for 2050, we performed a simple exercise involving wheat and rice, C3 species which might be expected to respond positively to successful bioengineering of the photosynthetic process. The assumptions underlying these projections are outlined in the legend of Fig. 1. Unsurprisingly, projected yields of the two crops, at current linear rates of yield increase, lag further and further behind demand as time from 2010 increases, the lag for rice being much worse than wheat (Fig. 1). By contrast, the sobering conclusion of these back-of-an-envelope speculations about the possible impact of improvements in photosynthesis, is that for rice, even the combination of a 30% step change in yield, with a 15 y delay in producing a farmer-ready cultivar, and full adoption in 5 y would only serve to meet projected maximum (i.e., at 1.31% y^{-1} rate of increase) demand between 2028 and 2030 (Fig. 1). All other combinations, although leading to improved yield over time, would be insufficient to meet projected demand at any time from now to 2050. For wheat, and with one exception, the only combinations of factors leading to satisfaction of projected maximum demand over two or more decades were a 30% step change in yield and a 15 y delay in obtaining farmer-ready cultivars (Fig. 1). The exception was that a 30% step change after 30 y, combined with full adoption in 5 y is projected to meet demand between 2043 and 2048.

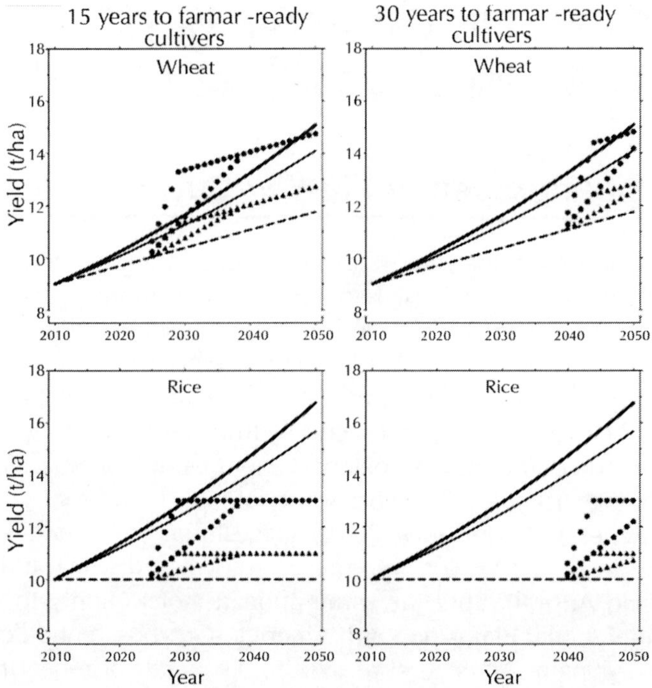

Figure 1: Possible 2010–2050 Yp trajectories for wheat and rice in relation to projected demand. Curves in each panel show required Yp assuming required compound rates of yield increase of 1.31% y^{-1} (upper curve) and 1.16% y^{-1} (lower curve) (cf. text Section 1). Continuous dashed line in each panel shows expected yield trajectory assuming current linear rates of yield increase (68 kg ha^{-1} y^{-1} for wheat [Mackay et al., 2011], 0 kg ha^{-1} y^{-1} for rice [Peng et al., 2010]) and initial (2010) values of 9 t ha^{-1} y^{-1} for wheat (Mackay et al., 2011) and 10 t ha^{-1} y^{-1} for rice (Bueno and Lafarge, 2009 and Peng et al., 2010). Superimposed on the above three lines are estimated trajectories resulting from introduction of farmer-ready cultivars incorporating large (30%, full circles) and moderate (10%, full triangles) increases in crop photosynthetic capacity (Zhu et al., 2010) after delays with respect to 2010 from idea to farmer-ready cultivars of 15-y (left-hand panels) or 30-y (right-hand panels). For each of the two step changes in crop photosynthetic capacity, two linear rates of adoption by farmers are considered: 100% adoption in 5 y (line closer to the *y*-axis) and 100% adoption in 15 y (line furthest from the *y*-axis). After complete adoption, current linear rates of yield progress are assumed to continue unchanged.

GENETIC IMPROVEMENT OF WATER-LIMITED YIELD POTENTIAL

Rates of Improvement in Yw: Current Status

The question of whether there has been genetic improvement in Yw can be viewed from at least two different viewpoints. Firstly, from sustained increases in yields over time, demonstrated either in side-by-side rainfed trials under standard management or in rainfed yield contests (e.g., Duvick and Cassman, 1999, Shearman et al., 2005 and Sadras and Lawson, 2011), and secondly from variety trials conducted under rainfed conditions over extended periods of time (e.g., Spink et al., 2009, Jaggard et al., 2010,Brisson et al., 2010 and Mackay et al., 2011). Fischer and Edmeades (2010) suggest that breeders have been successful in raising Yw for both maize in Iowa and wheat in France, the UK and Australia, but the same appears not to apply to rainfed wheat in the Great Plains of North America (Graybosch and Peterson, 2010) or in Spain (Acreche et al., 2008). Estimates of recent rates of progress in Yw for wheat in areas for which improvement has been demonstrated are close to 0.54% y^{-1} (compilation of side-by-side trial data for Australia by R. Fischer, 2009b), 0.87% y^{-1} in France (variety trial data, Brisson et al., 2010) and 0.50% y^{-1} (side-by-side trial data for South Australia, Sadras and Lawson, 2011). Duvick and Cassman (1999)provide data which suggests that improvement of Yw in rain-fed side-by-side comparisons for maize at optimum densities (i.e., those appropriate to the year of release of the hybrids) of 0.68% y^{-1} (1992) and 0.81% y^{-1} (1989) for maize in Iowa. Messina et al. (2009) provide data suggesting Yw for maize in a rain-fed target population of environments in the US was improving at 0.61% y^{-1} in 2007.

A second and complementary approach to the achievements described above using conventional breeding revolves around conscious attempts to improve Yw using physiological or molecular traits. A large number of putative physiological traits or black-box transgenes that might contribute to improved Yw have been proposed, and a very few of these have actually been demonstrated to be effective in the field for wheat (Richards, 2006 and refs. cited therein) and maize (Campos et al., 2004, Chimenti et al., 2002,Castiglioni et al., 2008 and Bänziger et al., 2006). Passioura, 1977 and Passioura, 2006 proposed that crop performance under water scarcity could be usefully examined within a framework that envisaged yield as the product of crop seasonal water use (WU), seasonal water use efficiency (WUE) and HI. He further argued that acceptance of putative traits that might contribute to greater yield under water scarcity should be conditional on the experimental demonstration that the trait or gene construct in question operated through alterations in one or more of WU, WUE and HI. Importantly, it has sometimes been possible to link the effects of some of these physiological traits to improvements in one or more of the elements of the Passioura identity, thus allowing a better appreciation of the mode of action of a given trait and its likely appropriateness to specific seasonal patterns of water availability and potential negative trade-offs. Thus, in maize, selection using anthesis-silking interval (ASI) and other traits improved HI (Edmeades et al., 1999, see Section 3.4), and higher osmotic adjustment led to greater WU and HI (Chimenti et al., 2002). In wheat, HI was improved by selection for narrow xylem vessels, WUE and HI were improved by selection using carbon isotope discrimination (CID) (Richards, 2006, see Section 3.4), and higher osmotic adjustment led to improved WU, WUE and HI (Morgan and Condon, 1986).

Yield loss mitigation under water-scarce conditions in rice (especially important for upland, rain-fed, rice) has received less attention in recent years than in wheat and maize. Nevertheless, a project involving a large number of diverse lines grown under managed stress environments has served to demonstrate the existence of significant genetic variability for yield loss mitigation under drought (e.g., Lafitte et al., 2006 and Lafitte et al., 2007), a fact which encourages further research on this issue. A large-effect QTL for yield under drought during the reproductive stage in rice has been reported in a study using a large number of F_3-derived lines originating in a cross between reportedly sensitive and tolerant lines (Bernier et al., 2007), but the significance of this is hard to discern because of the confounding associations between this QTL and time to flowering and, hence, the congruence between the timings of stress and flowering (cf.Reynolds et al., 2009).

Overall, important messages emerging from the work described above are that opportunities, both past and future, have existed or exist to increase Yw in wheat, maize and rice. In the past, these have been realised through selection for yield in rainfed multi-environment trials conducted within specific target environments (e.g., Duvick and Cassman, 1999, Shearman et al., 2005, Spink et al., 2009, Jaggard et al., 2010, Brisson et al., 2010 and Sadras and Lawson, 2011). Selecting for specific traits and yield in managed selection environments designed to mimic the main features of seasonal water availability has also served to effectively advance Yw (Campos et al., 2004) or to demonstrate the existence of exploitable genetic variability for water-scarce yield potential (Lafitte et al., 2006 and Lafitte et al., 2007). Managed stress environments also allow genetic advance for Yw to be quantified (e.g., Campos et al., 2004), serve to demonstrate the usefulness of particular traits (e.g., Campos et al., 2004, Chimenti et al., 2002 and Bänziger et al., 2006) and often facilitate the measurement of the three elements of the Passioura identity. Testing of germplasm bearing possible advantageous traits across a range of naturally occurring water-scarcity levels against check cultivars or sets of cultivars, as exemplified in Richards et al. (2007), also serves to demonstrate the usefulness of reputed traits.

The experience of inferred past gains in Yw through selection for yield in appropriate water-scarce environments, through exploitation of genetic variability generated by appropriate crossing programs, or via attempts to exploit physiological traits discovered through serendipity (e.g., anthesis-silking interval in maize, Campos et al., 2004) or following physiological reasoning (e.g., improved transpiration efficiency, TE, in wheat, Richards, 2006) all point to the fact gains are often fairly small in absolute terms (except under extreme drought) and relative rates of advance are not likely to be high. The yield gain, for example, from improved TE varied from 0% in the most favourable environment to 11% in the driest environment when contrasts were made in the same genetic backgrounds (Rebetzke et al., 2002), although when the contrast was made between high TE cv. Drysdale and low TE cv. Diamondbird yield gains varied from 0 to 35% in wet and dry environments respectively (Richards, 2006). For xylem vessel diameter the yield increased by around 7% in the driest environment and was neutral in the wettest environment (Richards, 2006). The *CsbP* transgene in maize (Castiglioni et al., 2008) provided a 33% increase in yield in the most stressful environment, tapering away toward neutrality in the better ones. These numbers also serve to show that gains will be strongly dependent on the intensity, duration and timing of the exposure to stress (see below). Another important lesson that has come out of all of these studies is that flowering time has a profound influence on yield under drought in all species and that when germplasm is being contrasted then flowering time of each line tested must be quantified so that its influence can be determined.

A great deal has been written about the use of transgenes or over-expressed native genomic regions as tools to improve Yw, but there are few examples in the literature (e.g., Castiglioni et al., 2008) in which adequate field proof has been provided. Even in this case, and given the often unique fit between the mode of action of an effective trait and nature of the target water availability regime, we still need to know how the trait works and what its effects on the elements of the Passioura identity are (Salekdeh et al., 2009).

Frameworks and Tools for Increasing Water-Limited Potential Yield

In water-limited environments, the problem of increasing Yw is even more complex than that of increasing Yp in unstressed environments. Seasonal patterns of crop-available water vary between environments and especially between years within environments (e.g., Chapman, 2008). Understanding the resulting genotype by environment interactions is essential to both the design of effective selection strategies for genetic improvement of yield and for correct interpretation of results obtained in multienvironment trials. A further dimension to this issue is that not all attributes that might contribute to yield loss mitigation under water scarcity are equally useful in all water-scarce environments (e.g., Richards, 2006). Additionally, the genetic architecture of the crop may condition the selection strategy that could lead to the optimum combination of genes or traits for a given environment (e.g., Hammer et al., 2009, Messina et al., 2009,Chenu et al., 2009 and Messina et al., 2011). And last, but unfortunately not least, the disconnect between many attempts to isolate genes or genomic regions that might be of interest as sources of yield loss mitigation under water scarcity and the realities of crops growing in water-scarce fields has led, in too many cases, to a focus on plant survival rather than on crop yield loss mitigation (e.g., Sinclair et al., 2004 and Passioura, 2010) as an indicator of tolerance to water scarcity.

The Passioura framework has gained a great deal of acceptance among breeders and crop ecophysiologists working in water-scarce environments (Richards, 2006) and, more recently, has been recommended in the context of molecular breeding (Salekdeh et al., 2009). Crop simulation models, combined with soil specifications and extended climatic records (e.g., Chapman, 2008) can provide a robust quantitative biophysical diagnosis that will, in combination with the Passioura framework, contribute to the understanding of crop responses to water-scarcity and help in the design of breeding strategies (including the use of managed water supply regimes, Campos et al., 2004) appropriate to given target environments. It seems highly probable that progress in Yw can be achieved using these tools and framework, but the necessary investment in time, experimentation and research facilities for full implementation of this strategy should not be underestimated.

In the context of opportunities and limitations for genetic improvement of Yw, a set of recent papers (Hammer et al., 2005, Messina et al., 2009, Messina et al., 2011 and Chenu et al., 2009), based on the combined use of process and genetic/breeding models, deserve special mention. They aim to show how the combination of these two types of models, given information about QTLs or genes linked to potentially interesting traits, can help breeders navigate complex yield-trait performance landscapes, break away from the one-trait-at-a-time approach, and mitigate uncertainties linked to GxExM interactions when using markers for multiple traits. The likely impact of this strategy on the rate of improvement in Yw is hard to assess at this time, either because some details of the work remain confidential (Messina et al., 2011), because the authors have had to use some plausible, but as yet unproven, relationships between processes such as silk extension and its genetic control (Chenu et al., 2009), or because the information requirements (especially, phenotyping) are high. In spite of these restrictions, these papers serve to illustrate the complexity (and likely slowness) of the way forward under water-limited conditions. Importantly, they also serve as a reality check for proposals aimed at increasing Yp (Spink et al., 2009) through breeding aimed at improving crop radiation capture, radiation use efficiency and HI.

Some private sector seed companies have relied on the use of transcription factors or chaperones (isolated from the crop species in question or originating in other species) to improve Yw (e.g., Nelson et al., 2007, Castiglioni et al., 2008 and Edgerton, 2009). Research reported in this latter group of papers highlights the potentially important issue that the physiological mode of action of these transcription factors is not reported, something which seriously hinders an appreciation of the water-stress regime for which they might prove effective (cf. Hammer et al., 2005). More informative as to physiological mode of action, but further away from testing in crop species, is the work on transgenes (e.g., Rivero et al., 2007) and transcription factors (e.g., Kasuga et al., 1999) that show some promise as sources of drought tolerance.

In summary, there are indications that genetic improvements in Yw have been achieved in some (albeit not in all) systems by selection in multi-environment rainfed trials or managed stress environments, and that there is or may be potential for further improvement. There have been limited demonstrated incremental gains in Yw using physiological traits and at least one transgenic approach seems to improve Yw in maize. The Passioura framework has been shown to be a useful tool in assessing the mode of action of some physiological traits, and crop simulation models (in combination with long-term climatic records and soil information) can provide very important insights to guide the selection process. New tools, such as the combined use of crop process and genetic models show promise as aids to the improvement process but, at the same time, signal the need for moderating expectations given the complex interactions between specific traits, a highly variable seasonal water supply, and the genetic architecture of the crop. Based on limited evidence, transgenes may provide a further avenue for improvement of Yw, but there needs to be a fundamental re-think of this issue in order to focus research on yield loss mitigation rather than plant survival or responses to unrealistic rates of stress development when searching for potentially useful genes.

The universe of suggested sources of genes/traits/genomic regions that might contribute to improved Yw is very large and impossible to encompass in this review. We have therefore decided to evaluate root system capacity to capture water as a very important trait (with relative impacts on Yw potentially as great as those of bioengineered photosynthesis on Yp) that merits further assessment (see Section 3.3). In addition, and because time (to 2050) is of the essence, we review the history of past successes in using physiological traits to improve Yw to provide some feeling for the time needed between the identification of a potentially useful trait and its possible release in farmer-ready cultivars (see Section 3.4).

Root System Capacity for Water Uptake: a Major Target for Yw Improvement?

It is recognised that below-ground processes have largely been neglected when considering ways to increase yield. This raises the important requirement to better understand root systems and what genetic opportunities may be available to breeding programs to overcome the various constraints. When soil constraints are remedied by management practices the root systems respond dramatically resulting in improved yields and this suggests that there are good prospects for directed genetic manipulation of root systems resulting in higher yields. The best examples are the application of both major and minor nutrients, seed dressings, rotations, liming and the success of these are widely known. In addition, changed management practices such as earlier sowing can also dramatically improve root system function for nutrient and water capture as the residency time for root systems is extended (Batten et al., 1999 and Lilley and Kirkegaard, 2011).

Almost all our crops are likely to be grown in soils which have some constraint. These may be soils which are too hard, too dry, too saline, too acid, limited by nutrients or have soil-borne diseases. Our knowledge of root system function and morphology in soils is still in its infancy and so targets for genetic modification are not always clear. Once identified it is expected that the development of effective selection methods will be challenging. There are of course exceptions to this and there have been some notable success stories. In wheat, for example, tolerance to acid soils is enhanced by a gene responsible for the release of malate from the roots (Delhaize and Ryan, 1995), for salinity the presence of sodium exclusion genes in the roots (Dvo ák et al., 1994 and Munns et al., 2012) and genes for boron tolerance (Jefferies et al., 2000). Several genes have been identified for tolerance to root cereal cyst nematode (Ogbonnaya et al., 2001) and root lesion nematode (Williams et al., 2002). For all of these examples molecular markers have been identified to improve selection in large breeding populations. Each of these examples is expected to increase nutrient and water uptake as a major limitation to these processes has been removed, and thereby improve both the yield of biomass and of grain.

The above examples are all associated with single identifiable genes that affect a specific root system characteristic. Most root system traits that affect root architecture and root function are likely to be multigenic and hence much more difficult to select for and less amenable to marker assisted selection. A good example of this is xylem vessel diameter in the seminal roots of wheat referred to above. Although this trait has a moderate heritability and can be selected for it is controlled by multiple genes each with a small effect. Lynch and colleagues (Lynch, 2007) have successfully selected for root traits that are multigenic to increase phosphorus acquisition. Evidence for other root system traits that are expected to increase yield are hard to find. Modelling suggests that deeper roots and more effective root systems have contributed to increased corn yields at higher planting densities in Iowa (Hammer et al., 2009), but Sinclair et al. (2010) found little indication of an equivalent advantage in the case of soybean, underlining the fact that traits need to be assessed in the context of species and cropping system. Lilley and Kirkegaard (2011)also show by modelling that breeding wheat with deeper roots in south-eastern Australia should increase water uptake and yield, but only in more favourable years. Manschadi et al. (2008) provide evidence that root angle of the seminal axes of successful wheat varieties in Australia is associated with the pattern of rainfall. They found that varieties with a higher angle of the seminal axis (less gravitropic) are more common in regions expected to receive their most important rainfall during the vegetative period whereas varieties with a narrow angle (more gravitropic), are likely to be adapted to regions that are reliant on deep water stored in the soil during grain filling. This also appears to be the case for nodal roots in sorghum (Mace et al., 2012) and rice (Kato et al., 2006).

The difficulties associated with measuring root systems in the field means that surrogate shoot traits which may reflect what is happening below-ground or are correlated with root growth can be invaluable. A novel example of this is placing a herbicide deep in the soil so as to select the genotypes which first show herbicide damage as these must have an ability to grow roots deep into the soil faster (Robertson et al., 1985). Other examples are shoot vigour which is correlated with root vigour and root length density (Liao et al., 2004 and Richards et al., 2007), reduced tillering in wheat – which is associated with more carbon allocation to roots (Duggan et al., 2005 and Richards et al., 2007), and a suite of above-ground traits may reflect the ability to capture deep water in the soil and so any one could be used for selection. The suites of traits are: canopy temperature, stomatal conductance, green leaf area or carbon isotope discrimination (Lopes and Reynolds, 2010 and Wasson et al., 2012). The use of surrogate above-ground traits is not, of course, free of risks. For example, higher canopy temperatures might be taken as an indication of a poor root water uptake capacity, but lower stomatal conductance would produce a similar signal but help spread crop water consumption across the season, ensuring enough water for grain filling (Richards, 2006 and Zaman-Allah et al., 2011). Whether the hidden half of crops will provide significant opportunities for improvement in the future is correct or not it would be perilous to neglect them at a time when the availability and cost-effectiveness of water and nutrients for crop production is coming under increased scrutiny.

Time from Idea to Farmer-Ready Cultivars. Some Case Histories

As with Yp (see Section 2.3), an issue that is often ignored in the debate about how and how fast genetic improvement might raise Yw, is the time required to progress from a putatively interesting idea or a discovery to farmer-ready cultivars (i.e., ones that meet all other requirements in terms of yield, product quality and disease resistance) that incorporate the trait of interest. Here, we make a first approximation to the issue by examining some past case histories (as distinct from the prospective analysis presented in Section 2.3) for the use of physiological traits for improved Yw but also to extend this by looking at the time required to incorporate semidwarf genes into wheat for favourable environments resulting in the Green Revolution and to incorporate herbicide and disease resistance in sunflower using molecular tools.

A great many physiological and molecular traits have been proposed as a means to improve water-limited potential yield, and only a few of these (Richards, 2006) have actually progressed to the proof-of-concept stage (defined here as a convincing demonstration of yield loss mitigation under field conditions representative of the water availability pattern that is being targeted), and even fewer have been incorporated into farmer-ready cultivars. Depending on the source of the trait of interest, the idea-to-proof-of-concept stage in this process can include identification of donors of the trait (or mutagenesis and screening), development of an understanding of the mode of inheritance, development of an effective selection method, and introgression of the trait into suitable test-bed lines using backcrosses and recurrent selection aided by physiological or molecular markers. Time required to complete this stage is a function of trait complexity, inheritance mode, screen simplicity, linkage drag, and money and efforts invested (serendipity has also been known to come into the equation).

One of the first conscious (as distinct from unconscious accumulation of favourable traits through selection for yield in rainfed environments) attempts to improve water-limited yield potential revolved around the rationing, by the plants, of the available soil water during the growing season in order to ensure sufficient residual water to support grain-filling (Passioura, 1972). Initially tested in large pots by restricting water uptake to a single seminal root, this idea evolved into the use of genetic variability in root xylem vessel diameter as a choke-point for water uptake. Variability for this trait and its heritability were explored byRichards and Passioura, 1981a and Richards and Passioura, 1981b; and the results of field proof-of-concept experiments (yield contrasts between two cultivars and their recurrently selected narrow-xylem-vessel backcrosses across a range of water availabilities) were published by Richards and Passioura (1989). In this example, time from idea to proof-of-concept was 17 years.

Sinclair et al. (1987) discovered that di-nitrogen fixation in soybean was particularly sensitive to drought, a finding which led to screening soybean lines for variability in sensitivity (Sall and Sinclair, 1991 and Sinclair et al., 2000), and the demonstration that yield under water scarcity improved in less-sensitive lines (Purcell and King, 1996). This was followed by a breeding program involving crosses between a low-sensitivity-to-drought donor and a high yielding soybean line and selection of two crosses which yielded better than the high yielding parent under water-scarce conditions (Sinclair et al., 2007) and the release of germplasm bearing the trait (Chen et al., 2007). Simulation modelling (Sinclair et al., 2010) suggests that this trait would produce yield benefits in more than 85% of year by region combinations in the US soybean-growing area. In this case, the time from the initial discovery to proof of concept in the field was 20 years.

Farquhar et al. (1982) demonstrated that discrimination against C^{13} (carbon isotope discrimination, CID) was associated with low transpiration efficiency (TE) of wheat, raising the possibility that CID could be used to distinguish between varieties for TE, as well as a trait which might be useful in mitigating early exhaustion of soil moisture reserves in environments subjected to late-season (for wheat) drought. Farquhar and Richards (1984) confirmed this linkage between CID and TE, leading to a breeding program that demonstrated that the trait had good heritability and that lines from backcrosses to existing commercial varieties had better yields under drought in field tests (Rebetzke et al., 2002). The first commercial variety bearing this trait was released in 2002 (Drysdale), with subsequent releases from 2004 onwards (Rees, Scout, and Envoy) (Richards, 2006). In this case, time from the idea to the first farmer-ready variety took 20 years.

Maize breeders and physiologists had long recognised that under drought conditions a delay in silking (i.e., the time between pollen shedding by the tassel and silk emergence on the spike) or anthesis-silking interval (ASI) was negatively associated with yield (e.g., DuPlessis and Dijkhuis, 1967 and Jensen, 1971) and there are indications that this trait was one of several used in private sector breeding for stress tolerance in the 1970s (G.O. Edmeades, personal communication). However, the first openly documented attempt to use ASI, combined with other traits such as leaf elongation, stay green and yield potential to form an explicit selection index, in a selection program conducted in managed drought stress environments is probably the one for tropical maize initiated in 1975 at CIMMYT by Ken Fischer and colleagues. This led to the first published results by Fischer et al. (1989) (i.e., an idea to proof of concept interval of 14 years), later expanded and confirmed on the basis of further work by Bolaños and Edmeades, 1993 and Bolaños and Edmeades, 1996. Although CIMMYT made improved tropical maize germplasm emerging from this selection process available, there does not appear to have been any documented cultivar releases. However, Bänziger and colleagues initiated a successful selection program for drought and N-stress tolerance in Zimbabwe in 1996 based on the CIMMYT-Mexico experience (reviewed inBänziger et al., 2006). This program led to the release, in 2000 (i.e., a proof-of-concept to farmer-ready cultivar interval of 11 years) of Zm521, which was rapidly adopted in the region. This release has been followed by other drought-tolerant OPV's and hybrids which have had an important regional acceptance.

The above examples of the time required to progress from an idea or discovery to farmer useable cultivars might overestimate these intervals because the research was not conducted on a full time basis. Nevertheless, each of the above traits are complex and controlled by many genes and this will almost certainly extend the time from discovery to adoption. It is likely that complex traits will be the norm for physiological traits that will improve Yw or Yp. It is therefore of interest to investigate the time taken to introduce a simply inherited trait into wheat that improves Yp for favourable growing conditions and the release of new varieties in south Asia. The best example is the introduction of dwarfing genes into wheat which, together with better rust resistance, has become known as the Green Revolution. Plant height has always been of major importance to wheat and rice breeders in favourable environments as it is the main way to reduce lodging. Until about the 1950s height had generally been considered a trait that is controlled by many genes each with a small effect. To manipulate it in breeding programs requires large populations compared to a trait that is controlled by a single gene that is easy to observe. Single genes responsible for plant height were not well documented in wheat in the 1950s. However, breeders in Japan were using single genes to reduce plant height in the 1800s without knowing it. The discovery of important wheat germplasm for height and its use in breeding outside Japan, and also the meteoric rise in adoption of semi-dwarf wheats has been documented by Dalrymple (1978). Briefly, Dr S.C. Salmon collected seed from a range of short wheats in Japan in 1946 and distributed them to seven breeding programs in the US. They were directly used in breeding by Dr Orville Vogel in Washington State University who was looking for ways to reduce straw length because of the increased use of artificial fertilisers and problems with lodging. Dr Norman Borlaug at the Rockefeller's International Wheat Improvement Project in Mexico had exhausted sources of dwarfness in breeding wheat and received seeds from Vogel in 1953. The first semi-dwarf variety was released in the US in 1961 and in Mexico in 1962 where a shuttle breeding program was used with 2 generations of selection each year. Seed of the first Mexican wheat arrived in India in 1962 and more in 1963 and the best of these were released as varieties in 1965. Thus, even for a simply inherited and easily selected trait, recognised as having very high value under favourable conditions, it took 15 years from germplasm identification to a new variety in the US and 12 years using a shuttle breeding program for new varieties in India.

We draw on two further examples, neither of which relate directly to cereals or Yw, but which may serve as minimum estimates of the idea-to-cultivar release intervals. Both relate to sunflower, a species for which current breeding technology allows 3–4 generations a year using embryo-rescue, and both relate to fairly simple traits amenable to marker-assisted selection for the trait in a context of genome-wide selection that ensures high value agronomic traits are maintained in selected material after the introgression of the trait of interest. In addition, in both cases proof of concept is fairly simple (i.e., yes or no) under glasshouse or small field plot conditions, thus avoiding the need for testing over several years and/or locations in the field to ensure adequate exploration of a range of environmental conditions. The first example relates to obtaining a herbicide-resistant mutant through mutagenesis (screening is performed by spraying mutagenised seedlings with herbicide) followed by development of a molecular marker and introgression of the trait into elite parental lines (Sala and Bulos, in press; Sala, personal communication). Time from formulation of the initial development protocol to successful isolation of a resistant mutant, determination of trait heritability and development of a marker was 5 years, and another 4 years passed before the first hybrids bearing the trait were released to farmers. The second example involves the introgression of resistance to verticillium wilt (caused by *V. dahliae*), an important biotic challenge for sunflower breeders (Creus et al., 2007). New resistance genes were identified using inoculations under controlled conditions, and molecular markers for these genes identified (Galella et al., 2012) and later used in the introgression process. The time required to progress from screening for sources of resistance to this and to other diseases of sunflower through to the development of markers and introgression of resistance traits into parental lines and release of new, resistant hybrids has taken between 8 and 14 years, depending on the complexity of the traits involved (Zambelli, personal communication).

Taken together, these case histories highlight the need to assign adequate weighting to the time required to progress traits of interest through development and testing and on to their incorporation into farmer-ready cultivars. We recognise that progress in the first three cases (narrow xylem vessels, drought tolerant di-nitrogen fixation, and CID) might have been faster if the research required had been the exclusive focus of the researchers involved and possibly if good molecular tools had been available to hasten introgression and production of converted cultivars. We also recognise that future developments in the techniques of molecular breeding and gene-to-phenotype models may help shorten times to commercial availability of new cultivars carrying the traits of interest. Nevertheless, the two sunflower cases, in which molecular tools were deployed and the number of generations per year that can be achieved is high, suggest that, even under these conditions, the process of trait identification, development and testing may still require a number of years.

CURRENT AND FUTURE BREEDING AND THE BREEDERS' TOOLBOX

Requirements for Successful Breeding

It is pertinent to consider what the nature of plant breeding is expected to be like in the future and whether the rate of breeding progress for Yp and Yw is likely to be enhanced as new enabling technologies become available and more widespread. This is important as the time involved from initial cross to varietal release in the majority of breeding programs where improved combinations of existing variation are being sought is often about 10 years. Conventional breeding will remain the mainstay of genetic advances in crops in the next decades. Thus, the fundamental principles and processes used today in breeding will almost certainly remain. However, actual breeding methodologies will continue to evolve and improve as they have over the last two decades and new technologies to enhance breeding processes will be implemented. Recent advances have centred on methodologies to improve selection efficiency and speed to market and they include the implementation of molecular markers for simply inherited traits, more efficient experimental designs and statistical data processing and more efficient data capture among other things (Knapp, 1998,Smith et al., 2005 and Kuchel et al., 2007). In addition, new thinking on gene to phenotype models for complex traits linked to a target population of environments (Hammer et al., 2005) and to yield-trait performance landscapes for agricultural environments (Messina et al., 2011) is also emerging which integrates the various – *omics* with information management and crop growth models. Genetic engineering, where new genes are introduced using transformation methods, will continue to develop as new genes and gene combinations of value are identified. However, these genes/traits will be integrated into, and be dependent upon, a conventional breeding program. In other words they will form part of the tool box of conventional breeders. Likewise it is expected that hybrid technologies will become more widespread in other crops as it has for rice in the last decade leading to further advances (Yang et al., 2007). The breeding of parent lines for hybrid production will also be dependent on a conventional breeding program.

The reason conventional breeding methods will remain the cornerstone of crop improvement is that they effectively and efficiently combine multiple traits together into a single new variety. Many traits contribute to yield and adaptation and these are most effectively selected in the target field environment where large populations can be grown. Selection for highly heritable traits critical for yield and adaptation such as height and flowering time can be made in early generations and many other traits can also be visually selected quickly and moderately effectively in the target environment. Such traits may be leaf development, vigour, maintenance of green leaves, canopy characteristics, synchronous reproductive development, size of reproductive organs, grain retention, lodging, pollen development, kernel size, colour, etc. Most of these are controlled by numerous genes most of which are unknown. Molecular methods have been of little value in these early segregating stages when very large populations are grown. In addition to selecting for these traits which may be required for yield and yield stability is the selection for resistance/tolerance to as many pests, diseases and soil constraints as possible. Selection for many of these can also be made visually in the target field environment although specialised environments for selection may sometimes be more effective. If they are simply inherited then molecular markers may be important for hard to measure traits (Kuchel et al., 2007) in specific populations. Grain characteristics for some crops are vital for consumer acceptance and also for the industrial processes used to manufacture acceptable food products, and these may also need to be selected. These are also complex genetically and where they are important they will inevitably slow breeding progress for yield. In wheat, for example, to be suitable to manufacturers of bread or noodles, the grain must give a high flour yield, and have desirable properties associated with water absorption, colour, colour stability, mixing time, extensibility, etc. as well as another suite of characters that measure end product features. Many genes, each of small effect, are associated with each of these characteristics (Li et al., 2009b).

It is in this overall breeding process that GM technologies are integrated so as to introduce new genes/traits into either hybrids or inbred lines for commercial release. Thus, to be competitive and to be accepted in the market place GM traits must be introduced into parent lines that are derived from conventional breeding programs and these then become integrated into a standard breeding program. Likewise, trait based breeding programs (e.g., Richards et al., 2010) where a trait responsible for improved yield, or provides tolerance to biotic or abiotic stress, or results in a more desirable end quality, will also need to be tested in conventional breeding programs to assess their competitiveness and potential in the market place.

The major technology changes heralded as revolutionary for crop breeding in the last 3 decades have been the use of marker assisted selection (MAS) and transformation technologies resulting in GM crops. Although MAS has been important for some simply inherited traits it has failed to be effective in breeding for complex traits associated with yield, tolerance to abiotic and biotic factors and grain quality (Bernardo, 2008 and Xu and Crouch, 2008). For complex traits multiple quantitative trait loci (QTL) each having a small effect are usually identified which makes MAS for complex traits almost impossible, particularly in variable environments, although gene-to-phenotype modelling (Hammer et al., 2005) may help to mitigate this problem. Also, populations used for QTL identification are often not representative of breeding populations and so marker implementation for complex traits may not be effective.

Dramatic changes in recent years in the cost of DNA sequence information, along with improved bioinformatic tools to interrogate the sequence information, promises to further drive and modify conventional breeding practices in the future. These changes will inevitably expand our molecular horizons beyond the use of single markers for genes with a large effect to complex traits which are affected by many genes each with a small effect. Genomic selection (GS) has been used very effectively in animal breeding (Hayes et al., 2009) and has considerable potential in crops (Zhong et al., 2009 and Jannink et al., 2010). SNP chips for the main crops are now available, the smallest being wheat with about 100,000 SNPs, and direct sequencing of germplasm is becoming widespread. These provide the opportunity to use all the polymorphic markers to predict crop performance and crop characteristics and this should enhance selection efficiencies. Genomic selection requires a training sub-population upon which accurate phenotypic data on yield and any other complex trait such as grain quality is obtained. Genomic estimated breeding values can be determined from the training population which is representative of the larger breeding population and these breeding values are then applied to the breeding population prior to any phenotypic evaluation (Zhong et al., 2009). It is emphasised that phenotypic data that is both accurate and repeatable will be critical so that the genomic information can be utilised effectively.

Underpinning the value of all molecular information for use in breeding is the importance and availability of fast and accurate phenotyping (Hammer et al., 2005, Finkel, 2009 and Richards et al., 2010). The phenotyping component is the bottleneck to further crop improvement as it is often difficult, expensive, time consuming, laborious and technically difficult and it is often destructive. In addition, in contrast to the genotype, the expression of the phenotype is subject to environmental and seasonal fluctuations and to genotype × environment interactions leading to a low repeatability and heritability. This is not unique to plants as the same problem limits progress in human medicine as well (Venter, 2011). An important key to successful phenotyping is to correctly design the screen used for phenotyping so that it mimics the target environment. Application of crop simulation models to the correct characterisation of stress regimes can be a powerful aid (e.g., Chapman, 2008), and the use of carefully defined managed stress environments played an important part in the success of the selection index for ASI, stay green and other traits in improving tropical maize Yw in Mexico and Zimbabwe (Fischer et al., 1989 and Bänziger et al., 2006).

Non-invasive technologies that measure plant form, function, aspects of metabolism and content, often called phenomics, has been touted as a way to overcome this bottleneck as measurements are often fast, repeatable and accurate and can be automated (Furbank and Tester, 2011). Nevertheless, a problem with much of the phenomics activities so far is the emphasis on controlled environments which, for most traits, rarely have relevance to the real environments in the field when the application is for breeding. However it is not unrealistic to think that non-invasive measurement of crop biomass, grain yield, grain nitrogen and grain moisture of an entire breeding trial should soon be possible, thereby improving the efficiency of the selection process.

GM crops that have been commercialised have been remarkably successful (James, 2011. Their success centres around the introduction of genes controlling the biochemistry of the plant to synthesise new chemical compounds (primarily Bt insect toxin) or to make the plants tolerate applied alien compounds (glyphosate tolerance). These alterations allow multiple crop species to be defended against a broad spectrum of insect pests and competition by weed species. There is a tendency to believe that these remarkable successes in GM technology will continue and will readily translate to new genes/traits which alter physiological or biochemical processes that influence yield or tolerance to abiotic stress. However, this translation is unlikely. Most traits we may wish to alter to increase yield or tolerance to abiotic stress are complex and it is unlikely that single genes will be important over a range of species and environmental conditions as we scale up from single plants to crop canopies and beyond where seasonal conditions are variable and unpredictable. The underlying gene networks and pathways are likely to dampen effects of single processes. Thus although GM technology will continue to be effective in altering end product quality and tolerance/resistance to biotic factors it is very unlikely these technologies will result in significant and widespread yield improvements directly as a result of a changed biochemistry. It is noteworthy that of the 39 commercial releases of GM traits in 2008 none directly targeted yield or abiotic stress; rather traits are for pest resistance/tolerance or product quality (Stein and Rodriguez-Cerezo, 2010).

Hindrances to Progress

The Green Revolution which commenced in the 1960s occurred at a time when most agricultural research was undertaken by publicly funded research organisations and germplasm and technology was freely exchanged between research partners and countries. During this time Intellectual Property (IP) was not a strong consideration in agricultural innovation. However, IP is now of considerable importance to both public and private research agencies to protect innovations required to increase food production and to transfer technologies and germplasm to collaborators and between countries. IP protection has become important in conventionally bred germplasm but its importance is exacerbated where GM technologies are involved. As IP has become more prominent in cultivar development there has been a corresponding decline in public investment in plant breeding but an increased private sector investment particularly in the developed countries. Thus, for the adoption of new varieties farmers in all countries need to adjust to the reality of the private sector having more control over seed distribution. Although this is unlikely to be an impediment in developed countries where hopefully it will speed up genetic progress, it could be an impediment in developing countries (Lea, 2008). Nevertheless, where multiple technologies owned by several agencies are used for cultivar development, as is likely for most traits/genes in GM technologies, then delays in cultivar release are likely because of regulatory issues. An example of this is the release of Golden Rice, which used GM technology to synthesise beta-carotene, a pre-cursor of vitamin A in the rice endosperm, which is expected to be of widespread health importance in rice dependent poor populations. Research on golden rice commenced in 1992 with the first field trials in 2004 and the expected release of cultivars to farmers in 2013 (Al-Babili and Beyer, 2005 and Potrykus, 2010). It is estimated that overcoming the regulatory hurdles related to GM as well as the many patents and organisations involved has delayed the release of the first cultivars by at least 10 years (Potrykus, 2010). The stringent regulatory requirements for the release of GM varieties is very costly and this is also expected to be a major impediment to, or in many cases prevent, the release of GM varieties. It is also likely to sideline public agencies as they will be unable to meet these costs on their own.

The time taken for the development and incorporation of new traits for yield or abiotic stress into cultivars by conventional breeders is also very slow as we outlined in Sections 2.3 and 3.4. In the overall research and development required for cultivar release it is likely that insufficient attention or funding is provided for the appropriate field evaluation of traits and the development of suitable germplasm for breeders. This is a major impediment to the adoption of a trait by breeders and hence delays or prevents cultivar release to farmers. For breeders to select for another trait in their breeding program they will first want proof of concept that the incorporation of the trait will result in an improvement in yield. This usually requires demonstration that in germplasm appropriate to their breeding programs (near-isogenic lines or large populations) there is an advantage when the trait is present. Once proven breeders will also want a fast and effective selection method to select for the trait and this may be either molecular markers or a fool-proof phenotyping method. Both of these can take time to develop and usually requires specialist skills (Richards et al., 2010). In the public sector proof of concept and germplasm development are often neglected, or do not happen as the required expertise for this is not available or there is the belief that the breeders should be responsible for it and so uptake of research results does not happen. Furthermore, funding agencies often do not appreciate the complexities and expertise required for adoption by breeders. This disconnect between pre-breeding and breeding may significantly delay or prevent uptake of new traits/technologies by breeders.

An often neglected factor that will be pivotal to achieving our food production targets is the availability of new cohorts of agricultural scientists trained in plant breeding, genetics, biotechnology, crop physiology and agronomy. Decreased government investment and support for agricultural sciences in our higher education systems have lead to reductions in crop research and shortages of trained staff in agribusiness, consulting, natural resource management and biosecurity. It is hoped that the global concern over food security and the efficient and equitable production and distribution of food, together with concerns over the maintenance of important ecosystems and natural resources, all of which are expressed in both the popular and scientific media, will inspire a new generation of young scientists.

CONCLUSIONS

Genetic variation in each of our major food crops has been successfully recombined by plant breeders over the last 100 years to make consistent yield gains. Genetic improvement together with better crop management resulted in the Green Revolution in the 1960s leading to dramatic improvements in yield, particularly under irrigation in developing countries, and in turn to food surpluses and to cheaper food in the developed world. However, in recent years there has been a frightening decline in global grain stocks and, together with adverse weather conditions in some regions, resulted in spikes in food prices (Koning et al., 2008) and social unrest in some countries. This as well as a sharp increase in fertiliser price and the concern over declining water resources available to agriculture provided the wake-up call required for society to realise that we may not be able to meet the global demand for food and feed grain by the middle of this century. To meet the demand for our major cereals, maize, wheat and rice, without increasing food prices then global harvests have to increase at an annual compound rate of between 1.16 and 1.31% y^{-1}. In examining results from long term studies that have evaluated elite genetic material and the best released varieties of maize, wheat and rice in well managed trials of historic sets of the best varieties or hybrids in both water unlimited and water-scarce environments we could find no examples that show *current* rates of increase of Yp or Yw that are close to the 1.16–1.31% y^{-1} required to satisfy projected demand for cereals in 2050. In fact a worrying number of cases of nil rates of increase of Yp or Yw were identified or suggested in various crop/environment combinations. These findings are broadly consistent with what has been found by others (e.g., Cassman et al., 2010) using subnational or national statistical records, but the management factor adds some uncertainty to the true significance of these observations.

Thus, in the challenge to progress Yp and Yw enough to meet projected demand for cereals by 2050, the timescales that have emerged are not a good basis for optimism. On the other hand, the revolution in biotechnology, primarily the low cost of DNA sequence information and advances in recombinant DNA technologies, together with fast and non-destructive phenotyping and computational wizardry, have provided some optimism (Phillips, 2010, Dunwell, 2011 and Furbank and Tester, 2011). However, whilst we can gain some useful insights thanks to these new tools, their impacts in terms of future yield gains and acceleration of the breeding process are still uncertain. There also remain unexploited opportunities for increasing Yp or Yw through targeted genetic improvement of the most important traits thought to limit yield. In addition, the processes to identify important traits for different regions have been improved by new tools to analyse seasonal variability and trait impact and these will provide a better focus. There are several routes to improve Yp or Yw and a number of candidate traits and genes are available. Here we identify photosynthetic capacity and root system function as two which we believe offers most promise in the future. A major consideration is that most of the important traits are genetically complex creating an added challenge. In addition, there prevails an unrealistic view of the time taken to incorporate new traits into varieties for farmers. In the case histories we considered of successful examples of trait integration into breeding the timescale from initial research to the release of new cultivars to farmers or to proof-of-concept in elite germplasm was around 20 years. Even in those cases in which introgression of fairly simple traits, governed by one or few genes, was backed by the full strength of current molecular breeding tools, this time span was of the order of 10 years. Variety adoption by farmers then occurs and this may be rapid in the case of blockbuster GM traits such as glyphosate resistance and pest resistance (Bt) or dwarfing genes in rice and wheat during the 1970s, but then longer for hybrid rice in China today. Times are expected to be much more gradual for other traits. Regulatory requirements and IP considerations are likely to add an extra level of uncertainty to the delivery time in the future.

Given the above, it is extremely hard to be optimistic that the current compound rates of progress in Yp and Yw will increase very much, if at all, in the near term even with the best methodologies and breeding programs. While this is obviously speculative, then so too is the optimistic (naïve) alternative. Also, the uncertainty and doubt over whether transgenic approaches will be effective (apart maybe from bioengineering photosynthesis) and the very long gestation period from the development of a new or existing trait to delivery in the form of new cultivars, leaves us with a very pessimistic view.

Whilst it is important to say that increased investment in research is going to be crucial just to maintain current rates of improvement, there is also an urgent need to focus on structural issues in the research, development and cultivar delivery chain, as well as on training a new generation of agricultural scientists. There remains a large gap between scientists close to the molecular level of organisation and the realities that breeders, agronomists and farmers have to face. Passioura (2010) strongly argues the need to change the way that research is generally organised and the important need to facilitate a culture of dialogue between scientists of different disciplines. This in turn is likely to focus attention on the most important and feasible approaches to crop genetic improvement and in turn redirect some research activity into more productive avenues. It should also enable a speedier transfer of traits to breeders. An increasing amount of cultivar development and delivery is through private companies and so private-public partnerships will be increasingly important to coordinate and to agree on mutually acceptable terms of engagement. The need for increased research investment and for structural/sociological change is clearly urgent if we are to meet the projected demand for food.

ROLE OF FUNDING SOURCES

The authors received no specific funding to support the preparation of this review.

ACKNOWLEDGMENTS

We thank Greg Edmeades, John Passioura, Carlos Sala, Tom Sinclair, and Andrés Zambelli for generously providing information, encouragement, opinion and critical comments without which Sect.3.4 would have been less true to life. Tony Fischer provided a critical appreciation of an earlier version of the manuscript. Mariano Oyarzábal assisted in estimating likely future impacts of step changes in Yp in wheat and rice as a function of delays in obtaining farmer-ready cultivars and variations in the rate of adoption of these. AJH is a member of CONICET, the National Scientific and Technological Research Council of Argentina.

REFERENCES

1. Acreche, M.M., Briceno-Félix, ˜ G., Martín Sánchez, J.A., Slafer, G.A., 2008. Physiological bases of genetic gains in Mediterranean bread wheat yield in Spain. Eur. J. Agric. 28, 162–170.
2. Aggarwal, P.K., Kropff, M.J., Cassman, K.G., ten Berge, H.F.M., 1997. Simulating genotypic strategies for increasing rice yield

potential in irrigated, tropical environments. Field Crops Res. 51, 5–17.

3. Al-Babili, S., Beyer, P., 2005. Golden rice: five years on the road–five years to go? Trends Plant Sci. 10, 565–573.
4. Bänziger, M., Setimela, P.S., Hodson, D., Vivek, B., 2006. Breeding for improved drought tolerance in maize adapted to southern Africa. Agric. Water Manage. 80, 212–224.
5. Batten, G.D., Fettell, N.A., Mead, J.A., Khan, M.A., 1999. Effect of sowing date on the uptake and utilization of phosphorus by wheat (cv. Osprey) grown in central New South Wales. Aust. J. Exp. Agric. 39, 161–170.
6. Bernardo, R., 2008. Molecular markers and selection for complex traits in plants: learning from the last 20 years. Crop Sci. 48, 1649–1664.
7. Bernier, J., Kumar, A., Ramaiah, V., Spaner, D., Atlin, G., 2007. A large-effect QTL for grain yield under reproductive-stage drought stress in upland rice. Crop Sci. 47, 507–518.
8. Berry, P.M., Spink, J.H., 2006. A physiological analysis of oilseed rape yields: past and future. J. Agric. Sci. (Camb.) 144, 381–392.
9. Bolanos, ˜ J., Edmeades, G.O., 1993. Eight cycles of selection for drought tolerance in lowland tropical maize. II. Responses in reproductive behavior. Field Crops Res. 31, 253–268.
10. Bolanos., ˜ J., Edmeades, G.O., 1996. The importance of the anthesis-silking interval in breeding for drought tolerance in tropical maize. Field Crops Res. 48, 65–80.
11. Brisson, N., Gate, P., Gouache, D., Charmet, G., Oury, F.-X., Huard, F., 2010. Why are wheat yields stagnating in Europe? A comprehensive data analysis for France. Field Crops Res. 119, 201–212.
12. Bruinsma, J.,2009. The resource outlook to 2050. By how much do land, water use and crop yields need to increase by 2050? In: Proc. FAO Expert Meeting on How to Feed the World in 2050, 24–26 June 2009. FAO, Rome (available at http://www.fao.org/wsfs/forum2050/background-documents/expert-papers/en/).
13. Bueno, C.S., Lafarge, T., 2009. Higher crop performance of rice hybrids than elite inbreds in the tropics. 1. Hybrids accumulate

more biomass during each phonological phase. Field Crops Res. 112, 229–237.

14. Bueno, C.S., Pasuquín, E., Tubana, ˜B., Lafarge, T., 2010. Improving sink regulation, and searching for promising traits associated with hybrids, as a key avenue to increase yield potential of the rice crop in the tropics. Field Crops Res. 118, 199–207.
15. Calderini, D.F., Slafer, G.A., 1998. Changes in yield and yield stability in wheat during the 20th century. Field Crops Res. 57, 335–347.
16. Campos, H., Cooper, M., Habben, J.E., Edmeades, G.O., Schussler, J.R., 2004. Improving drought tolerance in maize: a view from industry. Field Crops Res. 90, 19–34.
17. Cassman, K.G., Dobermann, A., Walters, D.T., Yang, H., 2003. Meeting cereal demand while protecting natural resources and improving environmental quality. Annu. Rev. Environ. Resour. 28, 315–358.
18. Cassman, K.G., Grassini, P., van Wart, J., 2010. Crop yield potential, yield trends and global food security in a changing climate. In: illel, D., Rosenzweig, C. (Eds.), Handbook of Climate Change and Agroecosystems: Impacts, Adaptation, and Mitigation. World Scientific, pp. 37–51.
19. Castiglioni, P., Warner, D., Bensen, R.J., Anstrom, D.C., Harrison, J., Stoecker, M., Kumar, G., Salvador, S., D'Ordine, R., Navarro, S., Back, S., Fernandes, M., Targolli, J., Dasgupta, S., Bonin, C., Luethy, M.H., Heard, J.E., 2008. Bacterial RNA
20. Chaperones confer abiotic stress tolerance in plants and improved grain yield in maize under water-limited conditions. Plant Physiol. 147, 446–455.
21. Chapman, S.C., 2008. Use of crop models to understand genotype by environment interactions for drought in real-world and simulated plant breeding trials. Euphytica 161, 195–208.
22. Chen, P., Sneller, C.H., Purcell, L.C., Sinclair, T.R., King, C.A., Ishibashi, T., 2007. Registration of soybean germplasm lines R01-416F and R01-581F for improved yield and nitrogen fixation under drought stress. J. Plant Growth Regul. 1, 166–167.
23. Chenu, K., Chapman, S.C., Tardieu, F., McLean, G., Welcker, C., Hammer, G.L., 2009. Simulating the yield impacts of organ-

level quantitative trait loci associated with drought response in maize: a "gene-to-phenotype" modeling approach. Genetics 183, 1507–1523.

24. Chimenti, C.A., Marcantonio, M., Hall, A.J., 2002. Divergent selection for osmotic adjustment results in improved drought tolerance in maize (Zea mays L.) in both early growth and flowering phases. Field Crops Res. 95, 305–315.
25. Cirera, X., Masset, E., 2010. Income distribution trends and future food demand. Philos. Trans. R. Soc. B 365, 2821–2834.
26. Cooper, P.J.M., Dimes, J., Rao, K.P.C., Shapiro, B., Shiferaw, B., Twomlow, S., 2008. Coping better with current climatic variability in the rain-fed farming systems of sub-Saharan Africa: an essential first step in adapting to future climate change? Agric. Ecosyst. Environ. 126, 24–35.
27. Creus, C., Bazzalo, M.E., Grondona, M., Andrade, F., León, A.J., 2007. Disease expression and ecophysiological yield components in sunflower isohybrids with and without Verticillium dahliae resistance. Crop Sci. 47, 703–708.
28. Dalrymple, D.G., 1978. Development and spread of high-yielding varieties of wheat and rice in the less developed nations. Foreign Agricultural Economic Report, USDA, No. 95. Washington DC, 134p.
29. Delhaize, E., Ryan, P.R., 1995. Aluminum toxicity and tolerance in plants. Plant Physiol.
30. 107, 315–321.
31. de laVega,A.J.,Chapman, S.C., 2010.Mega-environment differences affecting genetic progress for yield and relative value of component traits. Crop Sci. 50, 574–583.
32. de la Vega, A.J., DeLacy, I.H., Chapman, S.C., 2007. Progress over 20 years of sunflower breeding in central Argentina. Field Crops Res. 100, 61–72.
33. Duggan, B.L., Richards, R.A., van Herwaarden, A.F., 2005. Agronomic evaluation of a tiller inhibition gene (tin) in wheat II. Growth and partitioning of assimilate. Aust. J. Agric. Res. 56, 179–186.
34. Dunwell, J.M., 2011. Crop biotechnology: prospects and opportunities. J. Agric. Sci. (Camb). 149, 17–27.

35. DuPlessis, D.P., Dijkhuis, F.J., 1967. The influence oftime lag betweenpollenshedding and silking on the yield of maize. S. Afr. J. Agric. Sci. 10, 667–674.

36. Duvick, D.N., Cassman, D.N., 1999. Post–Green Revolution trends in yield potential of temperate maize in the North-Central United States. Crop Sci. 39, 1622–1630.

37. Dvorák, ˇ J., Noaman, M.M., Goyal, S., Gorham, J., 1994. Enhancement of the salt tolerance of Triticum turgidum L. by the Kna1 locus transferred from the Triticum aestivum L. chromosome 4D by homoeologous recombination. Theor. App. Genet. 87, 872–877.

38. Edgerton, M.D., 2009. Increasing crop productivity to meet global needs for feed, food, and fuel. Plant Physiol. 149, 7–13.

39. Edmeades, G.O., Bolanos, ˜ J., Chapman, S.C., Lafitte, H.R., Bänziger, M., 1999. Selection improves drought tolerance in tropical maize populations: I. Gains in biomass, grain yield, and harvest index. Crop Sci. 39, 1306–1315.

40. Ewert, F., Rounsevell, M.D.A., Reginster, I., Metzger, M.J., Leemans, R., 2005. Future scenarios of European agricultural land use I. Estimating changes in crop productivity Agric. Ecosyst. Environ. 107, 101–116.

41. Farquhar, G.D., O'Leary, M.H., Berry, J.A., 1982. On the relationship between carbon isotope discrimination and the intercellular carbon dioxide concentration in leaves. Aust. J. Plant Physiol. 9, 121–137.

42. Farquhar, G.D., Richards, R.A., 1984. Isotopic composition of plant carbon correlates with water-use efficiency of wheat genotypes.Aust.J. Plant Physiol. 11, 539–552.

43. Finkel, E., 2009. With 'Phenomics,' plant scientists hope to shift breeding into overdrive. Science 325, 380–381.

44. Fischer, G., 2009a. World food and agriculture to 2030/50: how do climate change and bioenergy alter the long-term outlook for food, agriculture and resource availability? In: Proc. FAO Expert Meeting on How to Feed the World in 2050, (available at http://www.fao.org/wsfs/forum2050/backgrounddocuments/expert-papers/en/).

45. Fischer, K.S., Edmeades, G.O., Johnson, E.C., 1989. Selection for the improvement of maize yield under moisture-deficits. Field Crops Res. 22, 227–243.

46. Fischer, R.A., 2009b. Farming systems of Australia: exploiting the synergy between genetic improvement and agronomy. In: Sadras, V.O., Calderini, D.F. (Eds.), Crop Physiology: Applications for Genetic Improvement and Agronomy. Academic Press, San Diego, CA, pp. 23–54.

47. Fischer, R.A., Edmeades, G.O., 2010. Breeding and cereal yield progress. Crop Sci. 50, S86–S98.

48. Fischer, R.A., Rees, D., Sayre, K.D., Lu, Z-M., Condon, A.G., Larqué-Saavedra, A., 1998. Wheat yield progress associated with higher stomatal conductance and photosynthetic rate and cooler canopies. Crop Sci. 38, 1467–1475.

49. Foulkes, M.J., Slafer, G.A., Davies, W.J., Berry, P.M., Sylvester-Bradley, R., Martre, P., Calderini, D.F., Giffiths, S., Reynolds, M.P., 2011. Raising yield potential of wheat. III. Optimizing partitioning to grain while maintaining lodging resistance. J. Exp. Bot. 62, 469–486.

50. Furbank, R.T., von Caemmerer, S., Sheehy, J., Edwards, G., 2009. C4 rice, a challenge for plant phenomics. Funct. Plant Biol. 36, 845–856.

51. Furbank, R.T., Tester, M., 2011. Phenomics–technologies to relieve the phenotyping bottleneck. Trends Plant Sci. 16, 635–644.

52. Galella, M.T., Bazzalo, M.E., Morata, M., Cimmino, C., Kaspar, M., Grondona, M., Reid, R., Zambelli, A., León, A., 2012. Pyramiding QTLs for Verticillium dahlia resistance. In: 18th International Sunflower Conf., Mar del Plata, Argentina, 27 February–1 March.

53. Grassini, P., Thorburn, J., Burr, C., Cassman, K.G., 2011. High-yield irrigated maize in the Western U.S. Corn Belt: I. On-farm yield, yield potential, and impact of agronomic practices. Field Crops Res. 120, 142–150.

54. Graybosch, R.A., Peterson, C.J., 2010. Genetic improvement in winter wheat yields in the Great Plains of North America, 1959–2008. Crop Sci. 50, 1882–1890. Godfray, H.C.J., Crute,

I.R., Haddad, L., Lawrence, D., Muir, J.F., Nisbett, N., Pretty, J., Robinson, S., Toulmin, C., Whiteley, R., 2010a. The future of the global food system. Philos. Trans. R. Soc. B 365, 2769–2777.

55. Godfray, H.C.J., Beddington, J.R., Crute, I.R., Haddad, L., Lawrence, D., Muir, J.F., Pretty, J., Robinson, S., Thomas, S.M., Toulmin, C., Whiteley, R., 2010b. Food security:the challenge of feeding 9 billion people. Science 327, 812–818.
56. Gornall, J., Betts, R., Burke, E., Clark, R., Camp, J., Willett, K., Wiltshire, A., 2010. Implications of climate change for agricultural productivity in the early twenty-first century. Philos. Trans. R. Soc. B 365, 2973–2989.
57. Hall, A.J., Sadras, V.O., 2009. Whither crop physiology? In: Sadras, V.O., Calderini, D.F. (Eds.), Crop Physiology: Applications for Genetiç Improvement and Agronomy. Elsevier, Amsterdam, pp. 545–570.
58. Hammer, G.L., Chapman, S.C., van Ooosterom, V., Podlich, D.W., 2005. Trait physiology and crop modelling as a framework to link phenotypic complexity to underlying genetic systems. Aust. J. Agric. Res. 56, 947–960.
59. Hammer, G.L., Dong, Z., McLean, G., Doherty,A., Messina, C., Schussler,J., Zinselmeier, C., Paszkiewicz, P., Cooper, M., 2009. Can changes in canopy and/or root system architecture explain historical maize yield trends in the U.S. Corn Belt? Crop Sci. 49, 299–312.
60. Hayes, B.J., Bowman, P.J., Chamberlain, A.J., Goddard, M.E., 2009. Genomic selection in dairy cattle: progress and challenges. J. Dairy Sci. 92, 433–443.
61. Hibberd, J.M., Sheehy, J.E., Langdale, J.A., 2008. Using C4 photosynthesis to increase the yield of rice—rationale and feasibility. Curr. Opin. Plant Biol. 11, 228–231.
62. Horie, T., Matsuura, S., Takai, T., Kuwasaki, K., Ohsumi, A., Shiraiwa, T., 2006. Genotypic difference in canopy diffusive conductance measured by a new remote-sensing method and its association with the difference in rice yield potential. Plant Cell Environ. 29, 653–660.
63. Jaggard, K.W., Qi, A., Eric, S., Ober, E.S., 2010. Possible changes to arable crop yields by 2050. Philos. Trans. R. Soc. B 365, 2835–2851.

64. James, C., 2011. Global Status of Commercialized Biotech/GM Crops: 2010 ISAAA Brief No. 42. ISAAA, Ithaca, NY, ISBN: 978-1-892456-49-4.

65. Janaiah, A., Xie, F., 2010. Hybrid rice adoption in India: farm level impacts and challenges. IRRI Tech. Bull. no. 14. IRRI. Los Banos, ˜ Philippines, 20 p.

66. Jannink, J-L., Lorenz, A.J., Iwata, H., 2010. Genomic selection in plant breeding: from theory to practice. Brief. Funct. Genomics 9, 166–177.

67. Jefferies, S.P., Pallotta, M.A., Paull, J.G., Karakousis, A., Kretschmer, J.M., Manning, S., Islam, A.K.M.R., Langridge, P., Chalmers, K.J., 2000. Mapping and validation of chromosome regions conferring boron toxicity tolerance in wheat (Triticum aestivum). Theor. App. Gen. 101, 767–777.

68. Jensen, S.D., 1999. Breeding for drought and heat tolerance in corn. In: Sutherland, J.I., Falasca, R.J. (Eds.), Proc. 26th Annual Corn and Sorghum, Res. Conf. Chicago, December 14–16, 1971. ASTA, Washington, DC, pp. 198–208.

69. Kasuga,M., Liu, Q.,Miura, S.,Yamaguchi-Shinozaki,K., Shinozaki,K., 1999.Improving plant drought, salt, and freezing tolerance by gene transfer of a single stressinducible transcription factor. Nat. Biotechnol. 17, 287–291.

70. Kato, Y.,Abe, J., Kamoshita,A., Yamagishi, J., 2006. Genotypic variation in root growth angle inrice andits association withdeeprootdevelopmentinuplandfields with different water regimes. Plant Soil 287, 117–129.

71. Knapp, S.J., 1998. Marker assisted selection as a strategy for increasing the probability of selecting superior genotypes. Crop Sci. 38, 1164–1174.

72. Koemel, J.E., Guenzi, A.C., Carver, B.F., Payton, M.E., Morgan, G.H., Smith, E.L., 2004. Hybrid and pureline hard winter wheat yield and stability. Crop Sci. 44, 107–113.

73. Koning, N.B.J., van Ittersum, M.K., Becx, G.A., van Boekel, M.A.J.S., Brandenburg, W.A., van den Broek, J.A., Goudriaan, J., van Hofwegen, G., Jongeneel, R.A., Schiere, J.B., Smies, M., 2008. Long-term global availability of food: Continued abundance or new scarcity? Neth. J. Agric. Sci. 55, 229–292.

74. Kuchel, H., Fox, R., Reinheimer, J., Mosionek, L., Willey, N., Bariana, H., Jefferies, S., 2007. The successful application of a marker-assisted wheat breeding strategy. Mol. Breed. 20, 295–308.

75. Lafarge, T., Bueno, C.S., 2009. Higher crop performance of rice hybrids than elite inbreds in the tropics. 2. Does sink regulation, rather than sink size, play a major role? Field Crops Res. 114, 434–440.

76. Lafitte, H.R., Li, Z.K., Vijayakumar, C.H.M., Gao, Y.M., Shi, Y., Xu, J.L., Fu, B.Y., Yu, S.B., Ali, A.J., Domingo, J., Maghirang, R., Torres, R., Mackill, D., 2006. Improvement of rice drought tolerance through backcross breeding: evaluation of donors and selection in drought nurseries. Field Crops Res. 97, 77–86.

77. Lafitte, H.R., Yongsheng, G., Yan, S., Li, Z-K., 2007. Whole plant responses, key processes, and adaptation to drought stress:the case of rice. J. Exp. Bot. 58, 169–175.

78. Lea, D., 2008. The expansion and restructuring of intellectual property and its implications for the developing world. Ethical Theor. Moral Pract. 11, 37–60.

79. Li, J., Xin,Y., Yuan, L., 2009. Hybrid rice technology development: ensuring China's food security. IPFRI discussion paper 0918. IPFRI, Washington, DC, 28 p.

80. Li, Y., Song, Y., Zhou, R., Branlard, G., Jia, J., 2009b. Detection of QTLs for bread-making quality in wheat using a recombinant inbred line population. Plant Breed. 128, 235–243.

81. Liao, M., Fillery, I.R.P., Palta, J.A., 2004. Early vigorous growth is a major factor influencing nitrogen uptake in wheat. Funct. Plant Biol. 31, 121–129.

82. Lilley, J.M., Kirkegaard, J.A., 2011. Benefits of increased soil exploration by wheat roots. Field Crops Res. 122, 118–130.

83. Long, S.P., Zhu, X.G., Naidu, S.L., Ort, D.R., 2006. Can improvement in photosynthesis increase crop yields? Plant Cell Environ. 29, 315–330.

84. Lopes, M.S., Reynolds, M.B., 2010. Partitioning of assimilates to deeper roots is associated with cooler canopies and increased yield under drought in wheat. Funct. Plant Biol. 37, 147–156.

85. Luque, S.L., 2000. Bases ecofisiológicas de la ganancia genética en el rendimiento del maíz en la Argentina en los últimos 30 anos. ˜ MSc Thesis. University of Buenos Aires, 96 p.

86. Luque, S.F., Cirilo, A.G., Otegui, M.E., 2006. Genetic gains in grain yield and related physiological attributes in Argentine maize hybrids. Field Crops Res. 96, 383–397.

87. Lutz, A.G., KC, S., 2010. Dimensions of global population projections: what do we know about future population trends and structures? Philos. Trans. R. Soc. B 365, 2779–2791.

88. Lynch, J.P., 2007. Roots of the second green revolution. Aust. J. Bot. 55, 493–512. Mace, E.S., Singh, V., Van Oosterom, E.J., Hammer, G.L., Hunt, C.H., Jordan, D.R., 2012. QTL for nodal root angle in sorghum (Sorghum bicolour L. Moench) co-locate with QTL for traits associated with drought adaptation. Theor. Appl. Genet. 124, 97–109.

89. Mackay, I., Horwell, A., Garner, J., White, J., McKee, J., Philpott, H., 2011. Reanalyses of the historic series of UK variety trials to quantify the contributions of genetic and environmental factors to trends and variability in yield over time. Theor. Appl. Genet. 22, 225–238.

90. Manschadi, A.M., Hammer, G.L., Christopher, J.T., deVoil, P., 2008. Genotypic variation in seedling root architectural traits and implications for drought adaptation in wheat (Triticum aestivum L.). Plant Soil 303, 115–129.

91. Messina, C.D., Hammer, G., Dong, Z., Podlich, D., Cooper, M., 2009. Modeling crop improvement in a GxExM framework via gene-trait-phenotype relationships.

92. In: Sadras, V.O., Calderini, D.F. (Eds.), Crop Physiology: Applications for Genetic Improvement and Agronomy. Elsevier, Amsterdam, pp. 235–265.

93. Messina, C.D., Podlich, D., Dong, Z., Samples, M., Cooper, M., 2011. Yield–trait performance landscapes: from theory to application in breeding maize for drought tolerance. J. Exp. Bot. 62, 855–868.

94. Monteith, J.L., 1977. Climate and efficiency of crop production in Britain. Philos. Trans. R. Soc. B 281, 277–294.

95. Morgan, J.M., Condon, A.G., 1986. Water use, grain yield, and osmoregulation in wheat Aust. J. Plant Physiol. 13, 523–532.

96. Munns, R., James, R.A., Xu, B., Athman, A., Conn, S.J., Jordans, C., Byrt, C.S., Hare, R.A., Tyerman, S.D., Tester, M., Plett, D., Gilliham, M., 2012.Wheat grain yield on saline soils is improved by an ancestral transporter gene. Nat. Biotechnol. 30, 360–364.

97. Murchie, E.H., Pinto, M., Horton, P., 2008. Agriculture and the new challenges for photosynthesis research. New Phytol. 181, 532–552.

98. Nelson, D.E., Repetti, P.P., Adams, T.R., Creelman, R.A., Wu, J., Warner, D.C., Anstrom, D.C., Bensen, R.J., Castiglioni, P.P., Donnarummo, M.G., Hinchey, B.S., Kumimoto, R.W., Maszle, D.R., Canales, R.D., Krolikowski, K.A., Dotson, S.B., Gutterson, N., Ratcliffe, O.J., Heard, J.E., 2007. Plant nuclear factor Y(NF-Y) B subunits confer droughttolerance andleadto improvedcornyields on water-limitedacres. PNAS 104, 16450–16455.

99. Ogbonnaya, F.C., Subrahmanyam, N.C., Moullet, O., Majnik, J.D., Eagles, H.A., Brown, J.S., Eastwood, R.F.,Kollmorgen, J.,Appels, R., Lagudah, E.S., 2001. Diagnostic DNA markers for cereal cyst nematode resistance in bread wheat. Aust. J. Agric. Res. 52, 1367–1374.

100. Parry, M.A.J., Reynolds, M., Salvucci, M.E., Raines, C., Andralojc, P.J., Zhu, X.-G., Price, G.D., Condon, A.G., Furbank, R.T., 2011. Raising yield potential of wheat. II. Increasing photosynthetic capacity and efficiency. J. Exp. Bot. 62, 453–467.

101. Passioura, J.B., 1972. The effect of root geometry on the yield of wheat growing on stored water. Aust. J. Agric. Res. 23, 745–752.

102. Passioura, J.B., 1977. Grain yield, harvest index, and water use of wheat. J. Aust. Inst. Agric. Sci. 43, 117–120.

103. Passioura, J.B., 2006. Increasing crop productivity when water is scarce – from breeding to field management. Agric. Water Manage. 80, 176–196.

104. Passioura, J.B., 2010. Scaling up: the essence of effective agricultural research. Func. Plant Biol. 37, 585–591.

105. Peng, S., Cassman, K.G., Virmani, S.S., Sheehy, J., Khush, G.S., 1999. Yield potential trends of tropical rice since the release of

IR8 and the challenge of increasing rice yield potential. Crop Sci. 39, 1552–1559.

106. Peng, S., Laza, R.C., Visperas, R.M., Sanico, A.L., Cassman, K.G., Khush, G.S., 2000. Grain yield of rice cultivars and lines developed in the Philippines since 1966. Crop Sci. 40, 307–314.

107. Peng, S., Huang,J., Cassman,K.G., Laza,R.C.,Visperas,R.,Khush, G.S., 2010. The importance of maintenance breeding: a case study of the first miracle rice variety-IR8. Field Crops Res. 119, 342–347.

108. Phillips, R.L., 2010. Mobilizing science to break yield barriers. Crop Sci. 50, S99–S108. Potrykus, I., 2010. Regulation must be revolutionized. Nature 466, 561.

109. Purcell, L.C., King, C.A., 1996. Drought and nitrogen source effects on nitrogen nutrition, seed growth, and yield in soybean. J. Plant Nutr. 19, 969–993.

110. Rebetzke, G.J., Condon, A.G., Richards, R.A., Farquhar, G.D., 2002. Selection for reduced carbon isotope discrimination increases aerial biomass and grain yield of rainfed bread wheat. Crop Sci. 42, 739–745.

111. Reynolds, M., Manes, Y., Izanloo, A., Langridge, P., 2009. Phenotyping approaches for physiological breeding and gene discovery in wheat. Ann. Appl. Biol. 155, 309–320.

112. Reynolds, M.P., Delgado B, M.I., Gutiérrez-Rodríguez, M., Larqué-Saavedra, A., 2000. Photosynthesis of wheat in a warm, irrigated environment: I: genetic diversity and crop productivity. Field Crops Res. 66, 37–50.

113. Reynolds, M., Bonnett, D., Chapman, S.C., Furbank, R.T., Manès, Y., Mather, D.E., Parry, M.A.J., 2011. Raising yield potential of wheat. I. Overview of a consortium approach and breeding strategies. J. Exp. Bot. 62, 439–452.

114. Richards, R., 2006. Physiologicaltraits used in the breeding of new cultivars for water scarce environments. Agric. Water Manage. 80, 197–211.

115. Richards, R.A., Passioura, J.B., 1981a. Seminal root morphology and water use of wheat. I. Environmental effects. Crop Sci. 21, 249–252.

116. Richards, R.A., Passioura, J.B., 1981b. Seminal root morphology and water use of wheat. II. Genetic variation. Crop Sci. 21, 253–255.

117. Richards, R.A., Passioura, J.B., 1989. A breeding program to reduce the diameter of the major xylem vessel in the seminal roots of wheat and its effect on grain yield in rain-fed environments. Aust. J. Agric. Res. 40, 943–950.

118. Richards, R.A., Watt, M., Rebetzke, G.J., 2007. Physiological traits and cereal germplasm for sustainable agricultural systems. Euphytica 154, 409–425.

119. Richards, R.A., Rebetzke, G.J., Watt, M., Condon, A.G., Spielmeyer, W., Dolferus, R., 2010. Breeding for improved water productivity in temperate cereals: phenotyping, quantitative trait loci, markers and the selection environment. Funct. Plant Biol. 37, 85–97.

120. Rivero, R.M., Kojima, M., Gepstein, A., Sakakibara, H., Mittler, R., Gepstein, S., Blumwald, E., 2007. Delayed leaf senescence induces extreme droughttolerance in a flowering plant. PNAS 104, 19631–19636.

121. Robertson, B.M., Hall, A.E., Foster, K.W., 1985. A field technique for screening for genotypic differences in root growth. Crop Sci. 25, 1084–1090.

122. Sadras, V.O., Lawson, C., 2011. Genetic gain in yield and associated changes in phenotype, trait plasticity and competitive ability of South Australian wheat varieties released between 1958 and 2007. Crop Past. Sci. 62, 533–549.

123. Sala. C.A., Bulos, M. Inheritance and molecular characterization of broad range tolerance to herbicides targeting acetohydroxyacid synthase in sunflower. Theor. Appl. Genet., in press. http://dx.doi.org/10.1007/s00122-011-1710-9.

124. Satterthwaite, D., McGranahan, G., Tacoli, C., 2010. Urbanization and its implications for food and farming. Philos. Trans. R. Soc. B 365, 2809–2820.

125. Salekdeh, G.H., Reynolds, M., Bennett, J., Boyer, J., 2009. Conceptual framework for drought phenotyping during molecular breeding. Trends Plant Sci. 14, 488–496.

126. Sall, K., Sinclair, T.R., 1991. Soybean genotypic differences in sensitivity of symbiotic nitrogen fixation to soil dehydration. Plant Soil 133, 31–37.

127. Sayre, K.D., Rajaram, S., Fischer, R.A., 1997. Yield potential progress in short bread wheats in North West Mexico. Crop Sci. 37, 36–42.

128. Searchinger, T., Heimlich, R., Houghton, R.A., Dong, F., Elobeid, A., Fabiosa, J., Tokgoz, S., Hayes, D., Yu, T.-H., 2008. Use of U.S. croplands for biofuels increases greenhouses gases through emissions from land-use change. Science 319, 1238–1240.

129. Shearman, V.J., Sylvester-Bradley, R., Scott, R.K., Foulkes, M.J., 2005. Physiological processes associated with wheat yield progress in the UK. Crop Sci. 45, 175–185.

130. Sinclair, T.R., Muchow, R.C., Bennett, J.M., Hammond, L.C., 1987. Relative sensitivity ofnitrogenand biomass accumulationto droughtinfield-grownsoybean.Agron. J. 79, 986–991.

131. Sinclair, T.R., Purcell, L.C., Vadez, V., Serraj, R., King, C.A., Nelson, R., 2000. Identification of soybean genotypes with N2 fixation tolerance to water deficits. Crop Sci. 40, 1803–1809.

132. Sinclair,T.R.,Purcell,L.C.,Sneller,C.H.,2004.Croptransformation and the challenge to increase yield potential. Trends Plant Sci. 9, 1360–1385.

133. Sinclair, T.R., Purcell, L.C., King, C.A., Sneller, C.H., Chen, P., Vadez, V., 2007. Drought tolerance and yield increase of soybean resulting from improved symbiotic N2 fixation. Field Crops Res. 101, 68–71.

134. Sinclair, T.R., Messina, C.D., Beatty, A., Samples, M., 2010. Assessment across the United States of the benefits of altered soybean drought traits. Agron. J. 102, 475–482.

135. Smith, A.B., Cullis, B.R., Thomson, R., 2005. The analysis of crop cultivar breeding and evaluation trials: an overview of current mixed model approaches. J. Agric. Sci. 143, 449–462.

136. Smith, P., Gregory, P.J., van Vuuren, D., Obersteiner, M., Havlík, P., Rounsevell, M., Woods, J., Stehfest, E., Bellarby, J., 2010. Competition for land. Philos. Trans. R. Soc. B 365, 2941–2957.

137. Spink, J., Street, P., Sylvester-Bradley, R., Berry, P., 2009. The potential to increase productivity of wheat and oilseed rape

in the UK. Report to the Government Chief Scientific Adviser, 126 p. (available at: http://www.commercialfarmers. co.uk/ PotentialProductivity.pdf).

138. Stein, A.J., Rodriguez-Cerezo, E., 2010. International trade and the global pipeline of new GM crops. Nat. Biotechnol. 28, 23–25.

139. Strzepek, K., Boehlert, B., 2010. Competition for water for the food system. Philos. Trans. R. Soc. B 365, 2927–2940.

140. Sylvester-Bradley, R., Foulkes, J., Reynolds, M., 2005. Future wheat yields: evidence, theory and conjecture. In: Sylvester-Bradley, R., Wiseman, J. (Eds.), Yields of Farmed Species. Nottingham University Press, Nottingham, UK, pp. 233–260.

141. Tittonell, P., Shepherd, K.D., Vanlauwe, B., Giller, K.E., 2008. Unravelling the effects of soil and crop management on maize productivity in smallholder agricultural systems of western Kenya—an application of classification and regression tree analysis. Agric. Ecosyst. Environ. 123, 137–150.

142. Tollenaar, M., 1991. Physiological basis of genetic improvement of maize hybrids in Ontario from 1959 to 1988. Crop Sci. 31, 119–124.

143. Vara, A.M., 2004. Transgénicos en Argentina: más allá del boom de la soja. Rev. Iberoam. Ciencia Tecnol. Sociedad 3, 101–129.

144. Venter, J.C., 2011. The human genome at 10: successes and challenges. Science 331, 546–547.

145. Wasson, A.P., Richards, R.A., Chatrath, R., Misra, S.C., Sai Prasad, S.V., Saxena, D.C., Rebetzke, G.J.,Kirkegaard,J.A.,Christopher,J. ,Watt,M., 2012. Traits and selection strategies to improve root systems and water uptake in water-limited wheat crops. J. Exp. Bot. 63, 3485–3498.

146. Williams, K., Taylor, S., Bogacki, P., Pallotta, M., Bariana, H., Wallwork, H., 2002. Mapping of the root lesion nematode (Pratylenchus neglectus) resistance gene Rlnn1 in wheat. Theor. Appl. Genet. 104, 874–879.

147. Xiao, Y.G., Qian, Z.G., Wu, K., Liu, J.J., Xia, X.C., Ji, W.Q., He, Z.H., 2012. Genetic gains in grain yield and physiological traits of winter wheat in Shandong Province, China, from 1969 to 2006. Crop Sci. 52, 44–56.

148. Xu, Y., Crouch, J.H., 2008. Marker-assisted selection in plant breeding: from publications to practice. Crop Sci. 48, 391–407.

149. Yang, W., Peng, S., Laza, R.C., Visperas, R.M., Dionisio-Sese, M.L., 2007. Grain yield and yield attributes of new plant type and hybrid rice. Crop Sci. 47, 1393–1400.

150. Zaman-Allah, M., Jenkinson, D.M., Vadez, V., 2011. Chickpea genotypes contrasting for seed yield under terminal drought stress in the field differ for traits related to the control of water use. Funct. Plant Biol. 38, 270–281.

151. Zhang, Q., 2007. Strategies for developing green super rice. PNAS 104, 16402–16409.

152. Zhong, S., Dekkers, J.C.M., Fernando, R.L., Jannink, J-L., 2009. Factors affecting accuracy from genomic selection in populations derived from multiple inbred lines: a barley case study. Genetics 182, 355–364.

153. Zhou, Y., He, Z.H., Sui, X.X., Xia, X.C., Zhang, X.K., Zhang, G.S., 2007. Genetic improvement of grain yield and associated traits in the northern China winter wheat region from 1960 to 2000. Crop Sci. 47, 245–253.

154. Zhu, X-G., Long, S.P., Ort, D.R., 2010. Improving photosynthetic efficiency for greater yield. Annu. Rev. Plant Biol. 61, 235–261.

Chapter 3

Phenotypic Diversity of Farmers' Traditional Rice Varieties in the Philippines

Roel C. Rabara , Marilyn C. Ferrer, Celia L. Diaz, Ma. Cristina V. Newingham, and Gabriel O. Romero

Philippine Rice Research Institute, Maligaya, Science City of Muñoz, Nueva Ecija 3119, Philippines

ABSTRACT

Traditional rice varieties maintained and cultivated by farmers are likely sources of germplasm for breeding new rice varieties. They possess traits potentially adaptable to a wide range of abiotic and biotic stresses. Characterization of these germplasms is essential in rice breeding and provides valued information on developing new rice cultivars. In this study, 307 traditional rice varieties newly conserved at the PhilRice genebank were characterized to assess their phenotypic diversity using

57 morphological traits. Using the standardized Shannon-Weaver diversity index, phenotypic diversity indices averaged at 0.73 and 0.45 for quantitative and qualitative traits, respectively. Correlation analyses among agro-morphological traits showed a high positive correlation in some traits such as culm number and panicle number, flag leaf width and leaf blade width, grain width and caryopsis width. Cluster analysis separated the different varieties into various groups. Principal component analysis (PCA) showed that seven independent principal components accounted for 74.95% of the total variation. Component loadings for each principal component showed morphological characters, such as culm number, panicle number and caryopsis ratio that were among the phenotypic traits contributing positive projections in three principal components that explained 48% of variation. Analyses of results showed high diversity in major traits assessed in farmers' rice varieties. Based on plant height and maturity, 11 accessions could be potential donor parents in a rice breeding program. Future collection trips and characterization studies would further enrich diversity, in particular traits low in diversity, such as anthocyanin coloration, awn presence, awn color, culm habit, panicle type and panicle branching.

INTRODUCTION

Rice is a major crop in the Philippines with a production of 16.7 million MT ranking the Philippines eighth in the world [1]. It is a major calorie source for most Filipinos and a major income source for 12 million farmers and their families [2]. Rice production contributes 2.2% to the Gross Domestic Product (GDP) of the country [3]. It is also a culturally important crop to Filipinos as rice is featured in many festivals and rituals [4, 5]. The Executive Order No. 1061 in 1985 acknowledged the importance of rice to the country's economy and livelihood by creating the Philippine Rice Research Institute (PhilRice) mandated to lead the country's rice research and development programs [6]. In the Philippines, the growth of the rice sector is highly dependent on yield improvements, which can be achieved through breeding new varieties and developing and promoting yield-enhancing technologies [2].

Rice breeders are constantly engaged in developing new rice varieties with higher yield potential to enhance the actual yield obtained by farmers in the field. One approach in plant breeding, proposed as early as 1968 is new ideotypes development [7]. This "plant-type concept of breeding" resulted from pioneering studies showing close associations between yield and certain morphological characters in response to nitrogen application [8, 9]. In 1966, the selection for the semi-dwarf rice plant type led to the release of the first modern high yielding variety, IR8, which commenced the "green revolution" in Asia [8]. After 28 years of successful release of the IR8, a yield plateau was observed and prompted rice breeders to propose a new plant type (NPT) during the International Rice Research Institute (IRRI) strategic planning workshop in 1993 [8,10]. As such, morphological characters rather than physiological traits were considered for NPT rice because they were easy to distinguish in a breeding program [11].

Rice genetic resources are key components to breeding programs, and farmers have played important roles in contributing to rice diversity by developing and nurturing thousands of rice varieties for several years [12]. This vast wealth of rice germplasm including landraces and traditional varieties is a good source of important alleles to develop new rice varieties. These germplasms serve as the foundation of any rice breeding program because they are the source of important traits necessary for improving and developing new breeds of rice varieties [13]. Several reports have shown the utilization of rice landraces in developing new varieties. IR8, dubbed a miracle rice [14], was the product from crosses between two landraces: a semi dwarf rice Dee-geo-woo-gen and tall, vigorous rice Peta [15]. The submergence tolerance SUB1 QTL was identified from submergence tolerant rice landrace FR13A. Its identification and characterization led to successful introgression of the QTL to rice mega-varieties [16]. Recently, theNAL1 allele that was identified from tropical Japonica rice landrace Daringan significantly increased the yield of modern rice cultivars [17].

Characterization of rice germplasms increases its utility in any breeding program. The use of agro-morphological traits is the most common approach utilized to estimate relationships between genotypes [18]. This approach was employed to assess diversity on ancestral lines of improved rice varieties in the Philippines [13], the indigenous rice in Yunnan, China [19] and the rice landraces in Nepal [18]. The conservation and characterization of these genetic resources is a necessity not only for posterity, but also for utilization in different improvement programs such as breeding for improved yield and tolerance to various stresses. It is important to assess the diversity of these germplasms materials to provide insights in the diversity of these germplasms.

Thus, this study assessed the phenotypic diversity of new rice germplasm farmers' varieties conserved at the PhilRice genebank. Information generated from phenotyping these germplasms can be used as basis for future collection trips to augment diversity in the genebank collections as well as baseline information for utilization in rice breeding programs.

RESULTS AND DISCUSSION

Germplasm Characterization

Diversity in Qualitative Traits

Phenotyping is an important activity to evaluate the utilization of the germplasm collection in a genebank. In this study, 307 traditional rice varieties recently conserved at the PhilRice genebank were scored and measured using 39 qualitative and 18 quantitative morphological characters. These germplasms were comprised of 215 Indica, 89 Javanica and three Japonica varieties (appendix Table A1). Among the qualitative characters scored, ligule shape and culm kneeling ability were observed invariants. All the germplasm characterized had a two-cleft ligule shape and the culms had no kneeling ability (Table 1). Twenty of the qualitative traits scored were dominated by one character in each trait with a distribution ranging between 76%–95%. As a result, these twenty agronomic traits had low diversity indices ranging between 0.12–0.45. These were mainly awn-related characters such as presence, color, distribution, and type. Awn color and panicle were the lowest calculated indices (H' = 0.12) because 95% of the varieties scored had no awn and 92% had a medium length panicle type. Most of the varieties had thick culms and an erect culm habit. Moderately diverse traits were observed for 15 descriptors with indices ranging between 0.46–0.74. Most of these traits were inflorescence-related traits such as panicle and spikelet characters. Diversity in caryopsis pericarp color (seed coat color) (Figure 1) was also evident with all states being represented in the rice varieties evaluated. White seed coat color was the predominant state (50.5%), followed by red seed coat color (30.0%), variable purple (5.9%), light brown (4.6%) and purple (4.2%).

Four of the 39 traits scored had a high diversity with an average index of 0.87. Two of these traits were culm-related which assessed rice sturdiness during maturity and harvest. Although the predominant character was intermediate lodging resistance, 37% of the rice varieties had strong to very strong lodging resistance at the mature stage. The endosperm type trait had the highest calculated diversity index of 0.99. The reason for this is that all the endosperm type descriptors (1 = non-glutinous, 2 = Intermediate, 3 = glutinous) were identified in the characterized germplasm.

Table 1: Qualitative traits showing the predominant state observed, distribution (%) and the calculated Shannon diversity indices (H') for each descriptor scored

Descriptor	**Predominant State**	**%**	**States Observed**	**H' Index**
Invariant				
Ligule Shape	2-ccenter	100.00	1	0.00
Culm Kneeing Ability	Absent	100.00	1	0.00
Low diversity				
Awn Color (Late)	Awnless	95.11	4	0.12
Panicle Type	Medium (~25 cm)	91.86	3	0.12
Awn Color (Early)	Awnless	91.53	6	0.18
Awn Distribution	Awnless	91.53	5	0.20
Culm Diameter Type	Thick	96.42	2	0.22
Awn Type	Awnless	91.21	5	0.22
Panicle Secondary Branching	Sparse	89.58	3	0.25
Culm Habit/Angle	Erect (<15°)	85.67	4	0.26
Sterile Lemma Type	Medium	62.54	3	0.27
Culm Anthocyanin Coloration on Nodes	Absent	90.55	4	0.28
Awn Presence	Absent	91.53	3	0.29
Leaf Blade Pubescence	Intermediate	91.86	3	0.29
Flag Leaf Attitude/Angle	Erect	83.71	4	0.30
Leaf Blade Attitude	Erect	89.90	3	0.31

Sterile Lemma Color	Straw	88.27	4	0.32
Panicle Attitude	Drooping	84.36	5	0.37
Stigma Color	White	77.85	4	0.41
Auricle Color	Whitish	75.90	6	0.45
Moderate diversity				
Lemma & Palea Color (Late Observation)	Straw	66.45	8	0.46
Panicle Attitude of Main Axis	Slightly drooping	70.03	3	0.49
Leaf Blade Length Type	Intermediate (~50 cm)	60.59	3	0.54
Culm Underlying Node Color	Green	73.29	3	0.56
Leaf Blade Width Type	Intermediate	65.15	2	0.59
Flag Leaf Attitude (Late Measurement)	Descending	71.01	4	0.63
Culm Length Type	Intermediate to long	36.81	6	0.63
Culm Number Type	Intermediate (~15 culms)	75.24	3	0.65
Late Lemma Apiculus Color	Straw	47.56	9	0.65
Lemma & Palea Pubescence	Short hairs	60.91	5	0.66
Panicle Exsertion	Well exerted	55.37	5	0.67
Caryopsis Pericarp Color (Seed Coat Color)	White	50.49	7	0.68
Caryopsis Shape	Long spindle-shaped	43.97	5	0.71
Productivity	Intermediate	70.03	3	0.71
Apiculus Shape	Curved	78.83	2	0.74
High diversity				
Early Lemma Apiculus Color	Straw	26.71	8	0.79
Culm Lodging Resistance	Intermediate	42.02	5	0.84
Culm Strength	Intermediate	41.04	5	0.84
Endosperm Type	Non-glutinous	41.37	3	0.99
Average diversity				0.45

Overall, the diversity in qualitative traits was low with an average index of 0.45. Several traits that were classified as low diversity might be prioritized in future collection trips to enhance their diversity in our genebank. Although not all these traits are linked to yield, some traits such as panicle and culm diameter types are useful parameters for improving yield. Wu et al., have shown that large culm rice varieties have a higher number of grains per panicle and a longer spike length [20].

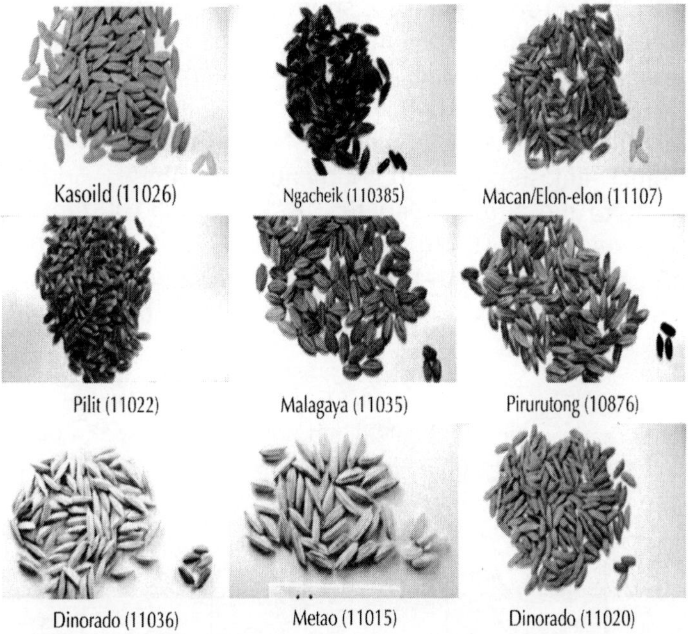

Figure 1: Diversity in grain color and caryopsis pericarp color of the different traditional rice varieties screened. Numbers in parenthesis are the collection numbers of the rice germplasm.

Diversity in Quantitative Traits

The 307 varieties showed diverse phenotypes in terms of plant height, leaf blade, flag leaf culm number and panicle number, among others. Table 2 sums up the quantitative morphological characters showing the highest and lowest values measured for each character. Most of the traits had moderate (5) to high (12) diversity indices. Awn length was the only trait that showed low diversity (H' = 0.02). This could be attributed to only 14 varieties exhibiting awns in their grains. Traits with moderate diversity included caryopsis (length, width and ratio) and culm descriptors (diameter and number). Culm number could be associated with yield if all tillers produced inflorescence. Culm strength and culm lodging resistance had diversity indices of 0.84. Maturity of characterized germplasm ranged from 71 to 154 days. Rice variety

Inuway (CollNo. 10869) had the shortest maturity. This variety was collected in an upland ecosystem in the Aurora province. Plant height varied from 68 cm to 161 cm, observed in Kinakaw (CollNo. 10857) and Dinorado (CollNo. 11049), respectively. Dinorado is an upland variety popular in the Arakan Valley of North Cotabato known for its sweet aroma, pinkish grain and good eating quality [4]. Additionally, the majority of the farmers' varieties (82%) were >100 cm tall with an average height of 116 cm and a median of 117 cm contradicting breeders' preference of 90–100 cm tall rice varieties among other characteristics to serve as potential donor for NPT breeding program [6].

Table 2: Quantitative descriptors and calculated Shannon-Weaver index (H′) of evaluated rice varieties, and calculated numbers enclosed in parenthesis are the collection numbers of the rice germplasm

Descriptors	H′	Min Trait Value	Variety	**Max Trait Value**	**Variety**	**Mean Trait Value (± Standard Deviation)**
Low Diversity						
Awn Length (mm)	0.02	1.98	Puchagwan (11241)	69.64	Burdagol (11083)	0.80 (6.01)
Moderate Diversity						
Caryopsis Length (mm)	0.64	4.37	Milagrosa (11102)	13.72	Binaka (10838)	6.49 (1.04)
Caryopsis Width (mm)	0.70	4.15	Kaimpas (11315)	1.28	Doriat Pula (10934)	2.23 (0.37)
Culm Diameter at Basal Internode (mm)	0.72	0.58	Fancy-1 (11105)	12.40	Speaker (11080)	8.06 (1.92)
Culm Number	0.73	5.20	Lubang (11341)	34.80	Rc 18 (Pula) (11312)	16.30 (4.48)
Caryopsis Length/Width Ratio Score	0.74	1.42	Kaimpas (11315)	5.12	Binaka (10838)	2.98 (0.60)
High Diversity						
Panicle Length (cm)	0.77	20.90	Pinarompong (10932)	38.80	Unoy (Umangan) (11245)	27.27 (2.67)
Maturity (day)	0.77	71.00	Inuway (10869)	154.00	18-Pula (11061); Galo (11205)	106.84 (35.69)

Grain Width (mm)	0.78	1.51	Duriat (10943)	6.56	Kipil (10837)	2.58 (0.53)
Panicle Number Per Plant	0.78	5.20	Lubang (11341)	34.80	Rc 18 (Pula) (11312)	16.26 (4.46)
Flag Leaf Length (cm)	0.79	23.80	Bolinao (11256)	72.80	Kinakaw (10857)	38.57 (7.56)
Leaf Blade Length (cm)	0.80	34.80	C-4 Dinorado (11317)	88.40	Lubang (11356)	58.36 (9.13)
Sterile Lemma Length (mm)	0.81	1.56	M10-2 (11086)	3.86	Minindoro (Pula) (10844)	2.47 (0.39)
100 Grain Weight (g)	0.81	1.10	Kinakaw (10857)	5.50	Binaka (10810)	2.45 (0.54)
Culm Length (cm)	0.82	67.80	Kinakaw (10857)	160.60	Dinorado (11049)	116.03 (15.67)
Grain Length (mm)	0.84	5.29	Dinorado (10860)	10.76	Binaka (10810)	8.48 (1.05)
Flag Leaf Width (cm)	0.85	1.10	Elon-Peta (11346)	2.60	Binisaya (10826)	1.95 (0.32)
Leaf Blade Width (cm)	0.85	0.96	Rc 18 (Pula) (11312)	2.48	Binisaya (10826)	1.64 (0.30)
Average Diversity	0.73					

Grain trait diversity was also observed in the germplasm (Figure 1). Of the 307 accessions, only 14 rice varieties had awns present on the grains with their lengths ranging from 2 mm to 70 mm with the variety Burdagol (CollNo. 11083) having the longest awn. The presence of awns is considered an important trait in rice domestication. Grains of wild rice have long awns that protect the grains from animal pilfering. Some reports suggest that the presence of awns in grains aids bird resistance in agricultural crops [21, 22]. Early studies in sorghum breeding have shown that varieties with long awns or that are strongly awned are more resistant to bird attacks than varieties with no awns [23]. Awned grains along those with few tillers and long panicles were found to be the characteristics of the bulu or Javanica group within the tropical Japonica varieties [24]. On the other hand, cultivated rice varieties have short awns allowing for easier harvesting than varieties with long awns [25]. Low tillering capability (three to four tillers) was one of the criteria used by IRRI rice breeders in selecting donor parents to be used in developing NPTs of rice [8]. Tillering ability among the 307 farmers' rice varieties ranged from five to 35 tillers. Rc 18-Pula (CollNo. 11312) showed the highest tillering ability among the farmers' varieties. It is a red-coated rice variety collected in the Bohol province and considered one of the popular varieties in the area based on Bertuso's survey [26]. This variety is a rice farmer's selection in 1997 [26] from his field planted with PSB Rc 18 (has a white seed coat), a modern rice variety released in 1994 for irrigated rice ecosystem [2]. Sturdy culm was another criterion used for donor parent selection for NPTs. The majority of the germplasms characterized (96%) had thick culms with a ≥5 mm culm diameter. However, when rice varieties were assessed for lodging resistance, only 37% showed strong or very strong lodging resistance. This could be attributed to the plant height, a major factor in lodging resistance in rice [27]. Having a short plant structure is currently the preferred trait for improving lodging resistance in rice [28].

In general, diversity in quantitative traits was moderate with an average index of 0.73. Nearly all the traits measured showed moderate to high diversity.

Correlation among Traits

Using Pearson's product-moment correlation, an analysis was done to assess the relationship among the morphological traits. It is useful to determine the relationship among the morphological traits since this information will be useful in the utilization of the germplasm as well in the collection of the germplasm based on the target traits. Several traits showed significant correlations ($r = 0.195$; $p < 0.05$) among each other. A heat map (Figure 2) was constructed to visualize the traits that had weak ($r \leq 0.35$), moderate ($r = 0.36$–0.67) and strong ($r = 0.68$–1.00) correlations [29]. An analysis showed that 89% of the trait combinations had weak correlations while 10% had moderate correlations. Only the correlation between the panicle number per plant and the culm number ($r = 0.998$) was strong. This showed that all tillers were productive tillers and able to bear inflorescence. The panicle number per plant ranged from five to 35. Sterile lemma length and 100-grain weight showed a moderate correlation ($r = 0.50$). This was expected since any increase in sterile lemma length would positively affect the grain weight. A high correlation was also observed between the flag leaf width and leaf blade width ($r = 0.59$) indicating that an increase in leaf blade width might also result in an increase of flag leaf width. A positive correlation was also observed between flag leaf width and grain width (r = 0.40). Flag leaves are important in grain filling, as 80% of the total carbohydrate stored in the grains is produced by the top two leaves in rice [30].

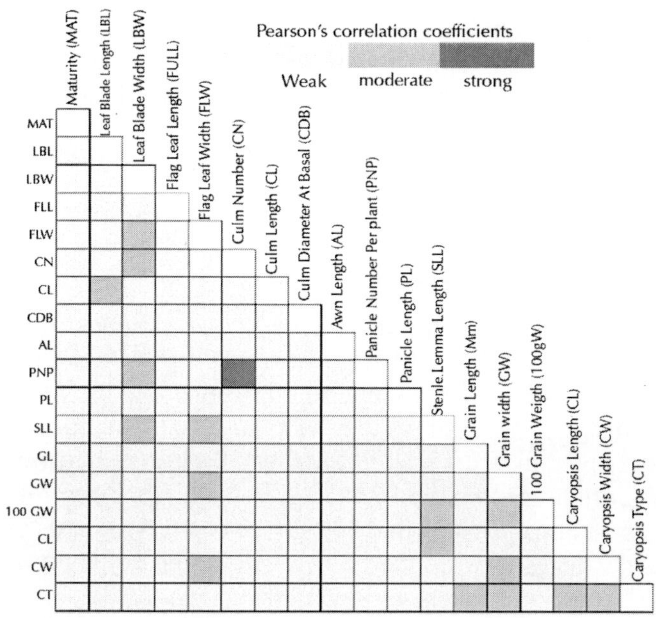

Figure 2: Heat map showing calculated Pearson's product-moment correlation coefficients among morphological traits measured in all germplasms screened. Correlation coefficients were classified as weak (r ≤ 0.35), moderate (r ≥ 0.36) and strong (r > 0.68) [29].

These characteristics are essential to rice breeders as it has been demonstrated that the flag leaf area increased grain yield by increasing the number of spikelets per panicle [31]. Flag leaves were reported to be the major source of phloem-delivered photoassimilates during the grain-filling stage in rice [32]. Previous studies have shown that cutting of flag leaves could result to up to 45% grain yield loss [33]. A recent review by Biswal and Kohli outlined the importance of flag leaf traits in cereal breeding for drought tolerance [34]. Flag leaf sheath is one of the main sources of carbohydrate for rice grain filling under drought condition [35].

Although the correlation analyses showed only one combination trait that had a strong correlation, there were 16 combination traits that had moderate correlations. These traits that had moderate to high correlations could be used as a basis for the utilization of these sets of germplasm for breeding purposes as well as for planning future collection trips targeting specific traits. Trait correlations can be used by breeders either to simultaneously improve correlated traits or reduce undesirable side effects when trying to improve only one of the correlated traits [36].

Cluster and Principal Component Analyses of Rice Germplasm

The relationship among the 307 farmers' rice varieties as revealed by Unweighted Pair Group Method with Arithmetic Mean (UPGMA) cluster analysis is shown in Figure 3. Truncating the tree at the Euclidean distance of 1.13 resulted in 24 clusters. In the truncated tree, 10 clusters had single accession, another 10 clusters had two to ten accessions; three clusters had 23–50 accessions and one big cluster had 137 accessions. The fact that 83% of the clusters formed contained one to a few accessions implied a diversity in the collection. Among the single-accession clusters, most accessions were collected from the Palawan and Kalinga provinces. The majority of these germplasms were of the Indica type while two accessions belonged to Javanica. These single-accession clusters were considered distinct from each other and the rest of the clusters. Cluster 3 (variety C-4 Dinorado; CollNo. 11317), for example, had similar traits to the accessions in Cluster 2 for most traits, such as leaf blade width type, flag leaf width, type, and erect flag leaf attitude at early stage, but differed by having very short leaf blades and horizontal flag leaf attitude at a late stage. Similarly, the variety Kinakaw (CollNo. 10857, Cluster 24) was distinct from the rest of the clusters because it had a short plant stature, the longest flag leaf length and the lightest 100-grain weight. Other single-accession clusters such as Cluster 21 (Burdagol CollNo. 11083), Cluster 22 (variety Chay-ot; CollNo. 11246) and Cluster 23 (variety Ifo; CollNo. 11255) were peculiar because of their long awns (30–70 mm). Most of the accessions (90%) had no awns while the rest had short awns (2–16 mm). Variety Benangkar (CollNo. 10923) was another single-accession cluster, which was characterized by the presence of purple lines in its culm nodes, which is a trait not very common among the rest of the accessions screened. Cluster 1, one of the four big clusters, was characterized as having a short plant stature (average height of 94.8 cm) compared to Clusters 2 (110.53 cm) and 4 (121.03 cm). Most of the accessions in Cluster 2 belonged to the Indica type (90%).

Overall, cluster analysis provided an insight into the diversity of the collections as shown by the number of clusters formed with one to 10 members when the dendrogram was truncated at 1.13 distance. A distinct variety was separated from the rest of the germplasm pool as exemplified by the single-accession clusters.

Principal component analysis (PCA) was employed to reduce the complexity of the data set while retaining the variation within the data set as far as possible [37]. The PCA resulted in 18 independent principal components that had a cumulative explained variance of 100% (Table 3). Following the Proportion of Variance Criterion [38], seven principal components (PCs) were retained that had a cumulative variance of 75%. The first component accounted for 22.5% of the total variation in the data set while the second and third principal components contributed 14.5% and 10.6%, respectively. Together, these three components could explain 47.6% of the total variation in the characterized rice germplasm. Analysis of the factor loadings of the characters in the retained PCs showed that phenotypic traits that contributed to yield showed high positive loadings in PC 1 (Table 4). These traits were culm number, panicle number per plant and caryopsis ratio score with factor loadings of 0.649, 0.651 and 0.529, respectively. These three morphological characters could have contributed to the maximum variability in PC 1 which explained 22.5% of the total variation in the data set. Among these three traits, only panicle number per plant was classified as high diversity while the other traits had moderate diversity indices. In PC 2, leaf blade length presented the highest factor loading of 0.482. This showed that leaf blade length was the major morphological character that contributed to the variation in PC 2 which explained 14.5% of the variation. In PC 3, plant height (culm length) showed a high loading of 0.454.

Figure 3: Dendrogram generated by cluster analysis of morphological characters using Unweighted Pair Group Method with Arithmetic Mean (UPGMA).

Table 3: Computed eigenvalues of the different principal components with corresponding proportion and cumulative explained variance

Components	Eigenvalue	**Explained Variance**	
		Percent	**Cumulative**
1	4.04	22.46	22.46
2	2.61	14.52	36.98
3	1.91	10.61	47.59
4	1.61	8.97	56.56
5	1.26	7.01	63.58
6	1.05	5.84	69.41
7	1.00	5.53	74.95
8	0.82	4.57	79.52
9	0.71	3.96	83.48
10	0.64	3.53	87.01
11	0.57	3.19	90.21
12	0.49	2.75	92.95
13	0.41	2.30	95.26
14	0.33	1.85	97.11
15	0.29	1.61	98.72
16	0.21	1.16	99.87
17	0.02	0.12	99.99
18	0.00	0.01	>100%

Table 4: Factor loadings (eigenvectors) for the different morphological characters for the principal components retained

Descriptors	Principal Components						
	PC 1	**PC 2**	**PC 3**	**PC 4**	**PC 5**	**PC 6**	**PC 7**
Maturity (day)	0.33	0.29	0.34	0.10	0.40	0.23	0.23
Leaf Blade Length (mm)	−0.11	0.48	0.53	0.14	0.12	−0.08	−0.31
Leaf Blade Width (mm)	−0.67	−0.21	−0.04	−0.26	−0.25	0.25	0.04
Flag Leaf Length (mm)	−0.17	0.21	0.43	0.46	−0.11	0.13	−0.26
Flag Leaf Width (mm)	−0.73	−0.12	−0.01	0.00	−0.14	0.21	0.21
Culm Number	0.65	−0.28	−0.36	0.52	−0.08	0.17	−0.09

Culm Length (cm)	−0.34	0.37	0.45	0.11	−0.14	0.33	0.02
Basal Culm Diameter (mm)	−0.33	−0.33	−0.01	0.15	−0.32	0.54	0.17
Awn Length (mm)	−0.11	−0.01	0.06	0.22	−0.54	−0.57	0.41
Panicle Number/plant	0.65	−0.28	−0.36	0.51	−0.08	0.18	−0.10
Panicle Length (cm)	−0.27	0.01	0.33	0.59	−0.27	−0.20	0.01
Sterile Lemma Length (mm)	−0.51	−0.67	0.10	−0.12	0.18	−0.03	−0.08
Grain Length (mm)	0.06	−0.47	0.37	0.26	0.46	−0.06	0.46
Grain Width (mm)	−0.66	0.13	−0.29	0.34	0.39	−0.09	0.10
100-Grain Weight (g)	−0.51	−0.46	−0.03	0.28	0.29	−0.06	−0.02
Caryopsis Length (mm)	−0.10	−0.73	0.31	−0.08	−0.05	−0.14	−0.43
Caryopsis Width (mm)	−0.72	0.08	−0.38	0.18	0.10	−0.13	−0.33
Caryopsis Length/ Width Ratio Score	0.53	−0.60	0.52	−0.18	−0.10	−0.02	−0.05

EXPERIMENTAL SECTION

Germplasm Characterization

A total of 307 rice germplasms conserved at the Philippine Rice Research Institute (PhilRice) genebank which were collected from different parts of the country (Figure 4) were used to evaluate the genetic diversity of farmers' traditional rice varieties and landraces. Characterization of the collected rice germplasms was done at the at the PhilRice Central Experimental Station, Maligaya, Science City of Munoz, Nueva Ecija (15°40′ N, 120°53′ E, 57.6 masl) during the wet cropping season (WS) in 2009. Prior to sowing, seeds were incubated in the oven set at 37 °C for 12 h to break inherent seed dormancy. Seeds were sown in a raised seedbed in the greenhouse and covered with coconut coir dust. Seven-day-old seedlings were transplanted into the field following a planting distance of 25 cm × 25 cm. A total of 20 plants per accession in three replicates were planted in the field for characterization. Agronomic characters were measured and scored following the Rice Descriptors [39]. A total of 18 quantitative and 39 qualitative characters were selected to score the germplasm collections.

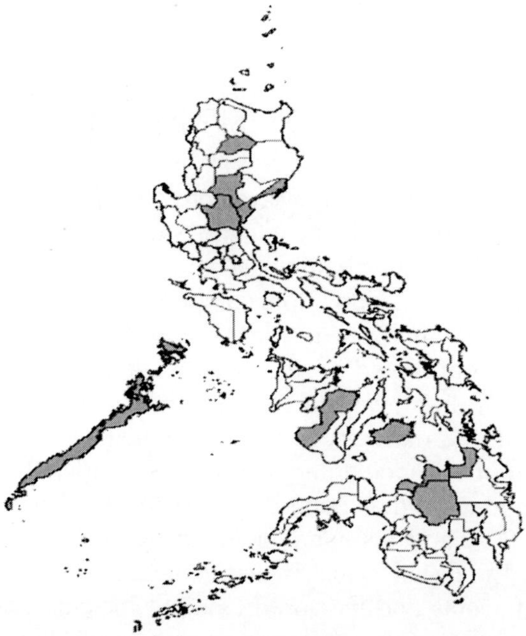

Figure 4: Philippine map showing the provinces (areas in green) where the rice germplasm were collected. The map was generated using DIVA-GIS ver.7.5.0 [40].

Data Analyses

Descriptive statistics was done using PROC UNIVARIATE procedure in SAS ver. 9.3 (SAS Institute, Cary, NC, USA [41]). Correlations among morphological traits measured were analyzed using PROC CORR procedure in SAS. Shannon-Weaver diversity index (H') was used to calculate the phenotypic diversity of the characterized farmers' varieties following the protocol used by Sotto and Rabara [42]. An arbitrary scale was adapted from Jamago and Cortes [43] to categorize the computed indices into maximum (H' = 1.00), high (H' = 0.76–0.99), moderate (H' = 0.46–0.75) and low diversity (0.01–0.45). Diversity indices of collected germplasm were calculated based on phenotypic frequency using standardized Shannon-Weaver Diversity index formula:

$$H' = -\sum pi\ (\log_2 pi)/\log_2 N$$

Wherein

Pi = frequency proportion of the descriptor state

N = number of states

The standardized Shannon-Weaver provided a constrained index between zero and one with the highest value indicating maximum abundance [44, 45]. Multivariate statistical analyses of characterization data were conducted using principal component (PCA) and cluster (CA) analyses. PCA was employed to identify the different morphological characters that contributed to the most variance in the measured variables. In PCA, the raw data were standardized and the distance matrix using the variance-covariance coefficients was computed. The Proportion of Variance criterion was used to identify the different principal components that contributed to the total variance in the dataset [23]. PCA and CA were done using NTSYSpc version 2.1 software [46]. The distance matrix was generated using the Euclidean Distance Coefficients and was used as input for clustering using the unweighted pair group of arithmetic means (UPGMA) method.

CONCLUSIONS

Phenotyping of germplasm materials is an important undertaking in genetic resource conservation to ensure efficient conservation management as well as its effective utilization especially in breeding programs. In this study, 307 rice varieties were characterized to assess their phenotypic diversity. Diversity analyses showed that 46% of the qualitative traits scored had low diversity indices compared to only 5% in quantitative traits. Overall, the rice germplasm showed moderate diversity based on quantitative characters (average index of 0.73). In contrast, the qualitative characters had a low diversity (average index of 0.45). In comparison to other landraces collections, the diversity in qualitative traits of our germplasm was higher than what Bajracharya et al., observed in the landraces from Nepal [18]. The complete qualitative data set of our collection is available in appendix Table A1.

In order to enrich the diversity of qualitative traits in our collection at the PhilRice genebank exploration trips may be conducted. Collection gaps should be identified in the genebank's germplasm collection and should be prioritized in future collection trips. An emphasis on farmers' varieties that have been planted in a community for several generations should be considered when planning for collection trips. It should also considered to look for diversity in qualitative characters that have shown low diversity indices such as presence of awns, panicle type, and culm habit type among other characters. Other factors that can be considered during collection of these rice germplasms are the characters that have high correlations with other characters. Collecting diversity for certain traits or characters could also lead to a high diversity if the characters are highly correlated.

The success of any conservation program could also be measured by the amount of which these genetic resources are being utilized. Data generated from characterization of these 307 rice varieties can be utilized as baseline information for the utilization of these germplasms for any rice breeding program such as breeding for ideoypes. For example, 37 accessions meets the plant height criterion (90–100 cm) for NPT among which 11 accessions also meets the maturity requirement (110–120 days). Short stature variety was one of the criteria to select a variety for use in a breeding program to address lodging. The optimal growth duration to achieve maximum yield is about 120 days [8]. A linear increase in total biomass had been observed when the growth duration was increased from 95 days to 135 days [47]. Efforts should be invested in promoting various stakeholders in rice production to utilize these germplasms. Effective utilization of these germplasms can be enhanced if these materials are fully characterized and evaluated for their potential use in breeding programs. Evaluation of these germplasms should be conducted to assess their potential as donor parents for the breeding of new varieties with improved responses to various abiotic and biotic stresses. Broadening of the genetic base through utilization of diverse germplasms in breeding for new rice varieties may be able to break the yield barrier that rice breeders are currently trying to address.

ACKNOWLEDGMENTS

We would like to acknowledge the assistance of Rodel Valdez, Edgar Vallejo, Daisy Villanueva, Jackie Reyes and Robert Punzal in field preparation, characterization and maintenance of the germplasm in the field. We also appreciate the assistance of Jennifer Jara-Rabara for editing the manuscript. This study was supported by PhilRice grant no. FEP02-030 to RCR.

AUTHOR CONTRIBUTIONS

Roel C. Rabara designed the experiment, analyzed the data and written the manuscript. Marilyn C. Ferrer conducted the phenotyping experiments, assisted in data analysis and contributed in writing the manuscript. Celia L. Diaz contributed in phenotyping the germplasm. Ma. Cristina V. Newingham contributed in data encoding and database management. Gabriel O. Romero contributed in writing the manuscript.

REFERENCES

1. FAOSTAT; Classic Version. Food and Agriculture Organization of the United Nations: Rome, Italy, 2013. Available online: http://faostat.fao.org/site/339/default.aspx (accessed on 5 August 2013).
2. Sebastian, L.S.; Alviola, P.A.; Francisco, S.R. Bridging the rice yield gap in the Philippines. In Bridging the Rice Yield Gap in the Asia-Pacific Region; Papademetriou, M.K., Dent, F.J., Herath, E.M., Eds.; FAO Regional Office for Asia and the Pacific: Bangkok, Thailand, 2000; p. 13.
3. Cororaton, C.B. Philippine Rice and Rural Poverty: An Impact Analysis of Market Reform Using CGE; International Food Policy Research Institute (IFPRI): Washington, DC, USA, 2006.
4. Reyes, L.C. Banking seeds. Rice Today **2012**, 11, 16–19.
5. Aguilar, F.V. Rice in the Filipino Diet and Culture; Philippine Institute for Development Studies: Makati, Philippines, 2005.
6. Establishing the Philippine Rice Research Institute (PRRI). Executive Order 1061; PRRI: Manila, Philippines, 1985.
7. Donald, C.M. The breeding of crop ideotypes. Euphytica **1968**, 17, 385–403.
8. Peng, S.; Khush, G.; Cassman, K. Evolution of the new plant ideotype for increased yield potential. In Breaking the Yield

Barrier Proceedings of a Workshop on Rice Yield Potential in Favourable Environments; International Rice Research Institute: Manila, Philippines, 1994.

9. Yoshida, S. Physiological aspects of grain yield. Annu. Rev. Plant Physiol. **1972**, 23, 437–464.
10. Tonini, A.; Cabrera, E. Opportunities for Global Rice Research in a Changing World; International Rice Research Institute: Los Baños, Philippines, 2011.
11. Peng, S.; Khush, G.S.; Virk, P.; Tang, Q.; Zou, Y. Progress in ideotype breeding to increase rice yield potential. Field Crops Res. **2008**, 108, 32–38.
12. Bellon, M.; Pham, J.; Jackson, M. Genetic conservation: A role for rice farmers. In Plant Genetic Conservation: The in situ Approach; Maxted, N., Ford-Lloyd, B., Hawkes, J., Eds.; Chapman and Hall: London, UK, 1997; pp. 263–289.
13. Caldo, R.; Sebastian, L.; Hernandez, J. Morphology-based genetic diversity analysis of ancestral lines of Philippine rice cultivars. Philipp. J. Crop Sci. **1996**, 21, 86–92.
14. Ronald, P. A case study of rice from traditional breeding to genomics. In The Role of Biotechnology in a Sustainable Food Supply; Popp, J.S., Jahn, M.M., Matlock, M.D., Kemper, N.P., Eds.; Cambridge University Press: New York, NY, USA, 2012; p. 10.
15. Hargrove, T.; Coffman, W.R. Breeding history. Rice Today **2006**, 5, 34–38.
16. Bailey-Serres, J.; Fukao, T.; Ronald, P.; Ismail, A.; Heuer, S.; Mackill, D. Submergence tolerant rice: SUB1's journey from landrace to modern cultivar. Rice **2010**, 3, 138–147.
17. Fujita, D.; Trijatmiko, K.R.; Tagle, A.G.; Sapasap, M.V.; Koide, Y.; Sasaki, K.; Tsakirpaloglou, N.; Gannaban, R.B.; Nishimura, T.; Yanagihara, S.; et al. NAL1 allele from a rice landrace greatly increases yield in modern Indica cultivars. Proc. Natl. Acad. Sci. USA **2013**, 110, 20431–20436.
18. Bajracharya, J.; Steele, K.A.; Jarvis, D.I.; Sthapit, B.R.; Witcombe, J.R. Rice landrace diversity in Nepal: Variability of agro-morphological traits and SSR markers in landraces from a high-altitude site. Field Crops Res. **2006**, 95, 327–335.

19. Yawen, Z.; Shiquan, S.; Zichao, L.; Zhongyi, Y.; Xiangkun, W.; Hongliang, Z.; Guosong, W. Ecogeographic and genetic diversity based on morphological characters of indigenous rice (Oryza sativa L.) in Yunnan, China. Genet. Resour. Crop Evol. **2003**, 50, 567–577.
20. Wu, L.-L.; Liu, Z.-L.; Wang, J.-M.; Zhou, C.-Y.; Chen, K.-M. Morphological, anatomical, and physiological characteristics involved in development of the large culm trait in rice. Aust. J. Crop Sci. **2011**, 5, 1356–1363.
21. Bullard, R.; York, J. Breeding for bird resistance in sorghum and maize. In Progress in Plant Breeding-1; Russell, G., Ed.; Butterworths: London, UK, 1985; pp. 193–222.
22. Bullard, R.W. Characteristics of Bird-resistance in Agricultural Crops; Denver Wildlife Research Center: Denver, CO, USA, 1988.
23. Jowett, D. Breeding bird-resistant sorghum in East Africa. Plant Breed. Abstr. **1967**, 37, 85.
24. Vaughan, D.A.; Lu, B.-R.; Tomooka, N. The evolving story of rice evolution. Plant Sci. **2008**, 174, 394–408.
25. Hu, G.; Zhang, D.; Pan, H.; Li, B.; Wu, J.; Zhou, X.; Zhang, Q.; Zhou, L.; Yao, G.; Li, J.; et al. Fine mapping of the awn gene on chromosome 4 in rice by association and linkage analyses. Chin. Sci. Bull. **2011**, 56, 835–839.
26. Bertuso, A. Farmers' Management of Rice Genetic Diversity: A Study on Enhancing Red Rices in Bohol, Philippines, in Department of Crop Science; Wageningen University: Wageningen, The Netherlands, 2000; p. 144.
27. Hitaka, H. Studies on the lodging of rice plants. Jpn. Agric. Res. Q. **1969**, 4, 1–6.
28. Ookawa, T.; Hobo, T.; Yano, M.; Murata, K.; Ando, T.; Miura, H.; Asano, K.; Ochiai, Y.; Ikeda, M.; Nishitani, R.; et al. New approach for rice improvement using a pleiotropic QTL gene for lodging resistance and yield. Nat. Commun. **2010**, 1, 1–11.
29. Taylor, R. Interpretation of the correlation coefficient: A basic review. J. Diagn. Med. Sonogr. **1990**, 6, 35–39.

30. Gladun, I.; Karpov, E. Production and partitioning of assimilates between the panicle and vegetative organs of rice after flowering. Russ. J. Plant Physiol. **1993**, 40, 629–633.
31. Yue, B.; Xue, W.Y.; Luo, L.J.; Xing, Y.Z. QTL analysis for flag leaf characteristics and their relationships with yield and yield traits in rice. Acta Genet. Sin. **2006**, 33, 824–832.
32. Narayanan, N.N.; Vasconcelos, M.W.; Grusak, M.A. Expression profiling of Oryza sativa metal homeostasis genes in different rice cultivars using a cDNA macroarray. Plant Physiol. Biochem. **2007**, 45, 277–286.
33. Abou-khalifa, A.A.B.; Misra, A.; Salem, A.E.-A.K. Effect of leaf cutting on physiological traits and yield of two rice cultivars. Afr. J. Plant Sci. **2008**, 2, 147–150.]
34. Biswal, A.; Kohli, A. Cereal flag leaf adaptations for grain yield under drought: Knowledge status and gaps. Mol. Breed. **2013**, 31, 749–766.
35. Garcia, A.; Dorado, M.; Perez, I.; Montilla, E. Effect of water deficit on the distribution of photoassimilates in rice plants (Oryza sativa L.). Interciencia **2010**, 35, 46–54.
36. Chen, Y.; Lubberstedt, T. Molecular basis of trait correlations. Trends Plant Sci. **2010**, 15, 454–461.
37. Ringnér, M. What is principal component analysis? Nat. Biotechnol. **2008**, 26, 303–304.
38. O'Rourke, N.; Hatcher, L. A Step-by-Step Approach to Using SAS for Factor Analysis and Structural Equation Modeling, 2nd ed.; SAS Institute: Cary, NC, USA, 2013.
39. Bioversity International, International Rice Research Institute and WARDA Africa Rice Center. Descriptors for Wild and Cultivated Rice (Oryza spp.); Bioversity International. International Rice Research Institute: Rome, Italy, 2007.
40. Hijmans, R.; Guarino, L.; Cruz, M.; Rojas, E. Computer tools for spatial analysis of plant genetic resources data: 1. DIVA-GIS. Plant Genet. Resour. Newsl. **2001**, 127, 15–19.
41. SAS Institute. I. What's New in SAS 9.3; SAS Institute: Cary, NC, USA, 2012.
42. Sotto, R.; Rabara, R.C. Morphological diversity of Musa balbisiana in the Philippines. J. Nat. Stud. **2007**, 6, 37–46.

43. Jamago, J.M.; Cortes, R.V. Seed diversity and utilization of the upland rice landraces and traditional varieties from selected areas in Bukidnon, Philippines. IAMURE Int. J. Ecol. Conserv. **2012**, 4, 112–130.

44. Pielou, E.C. An Introduction to Mathematical Ecology; John Wiley & Sons Inc.: Hoboken, NJ, USA, 1969.

45. Lexerød, N.L.; Eid, T. An evaluation of different diameter diversity indices based on criteria related to forest management planning. For. Ecol. Manag. **2006**, 222, 17–28.

46. Rohlf, F. NTSYSpc:Numerical Taxonomy System; Version 2.1. Exeter Publishing, Ltd.: Setauket, New York, NY, USA, 2002.

47. Akita, S. Improving yield potential in tropical rice. In Progressin Irrigated Rice Research; International Rice Research Institute: Manila, Philippines, 1989; pp. 4173.

Chapter 4

The Cysteine Protease–Cysteine Protease Inhibitor System Explored in Soybean Nodule Development

Barend Juan Vorster[1], Urte Schlüter[2], Magdeleen du Plessis[1], Stefan van Wyk[1], Matome Eugene Makgopa[2], Ignatious Ncube[3], Marian Dorcas Quain[4], Karl Kunert[2], and Christine Helen Foyer[4]

[1]Department of Plant Production and Soil Science, Forestry and Agricultural Biotechnology Institute, University of Pretoria, Pretoria 0002, South Africa

[2]Department of Plant Science, Forestry and Agricultural Biotechnology Institute, University of Pretoria, Pretoria 0002, South Africa

[3]Biotechnology Unit School of Molecular and Life Sciences, University of Limpopo, Sovenga 0727, South Africa

[4]Center for Plant Sciences, Faculty of Biology, University of Leeds, Leeds, LS2 9JT, UK

ABSTRACT

Almost all protease families have been associated with plant development, particularly senescence, which is the final developmental stage of every organ before cell death. Proteolysis remobilizes and recycles nitrogen from senescent organs that is required, for example, seed development. Senescence-associated expression of proteases has recently been characterized using large-scale gene expression analysis seeking to identify and characterize senescence-related genes. Increasing activities of proteolytic enzymes, particularly cysteine proteases, are observed during the senescence of legume nodules, in which a symbiotic relationship between the host plant and bacteria (*Rhizobia*) facilitate the fixation of atmospheric nitrogen. It is generally considered that cysteine proteases are compartmentalized to prevent uncontrolled proteolysis in nitrogen-fixing nodules. In addition, the activities of cysteine proteases are regulated by endogenous cysteine protease inhibitors called cystatins. These small proteins form reversible complexes with cysteine proteases, leading to inactivation. However, very little is currently known about how the cysteine protease-cysteine protease inhibitor (cystatin) system is regulated during nodule development. Moreover, our current understanding of the expression and functions of proteases and protease inhibitors in nodules is fragmented. To address this issue, we have summarized the current knowledge and techniques used for studying proteases and their inhibitors including the application of "omics" tools, with a particular focus on changes in the cysteine protease-cystatin system during nodule development.

INTRODUCTION

Nitrogen deficiency is a widespread problem in many soils requiring the use of nitrogen fertilizers to increase crop productivity. Unfortunately, the widespread use of relatively expensive chemical fertilizers is beyond the reach of most farmers in developing countries. However, a less expensive nitrogen source is available to farmers in the form of the fixation of atmospheric nitrogen by root nodule bacteria (rhizobia) in food and forage legumes [1]. The energy required to drive atmospheric nitrogen reduction in nodules is produced from sucrose that originates from photosynthesis. Sucrose is translocated from the leaf to the nodule. In return, the symbiotic nitrogen fixation activities of the nodules provide the plant with organic nitrogen. Symbiotic nitrogen assimilation not only spares existing soil nitrogen supplies from depletion, but it acts as a mechanism of green soil fertilization that improves the organic content of soil and so makes a valuable contribution to sustainable agriculture [2]. Improving our current understanding of the factors that limit the effective establishment and duration of symbiosis is key to maximizing legume production by resource-poor farmers, especially in soils with low nitrogen. Some legumes are more achieve symbiotic nitrogen assimilation more effectively than others. For example, grain legumes, such as peanuts, cowpeas, soybeans, and also fava beans have high rates of nitrogen fixation, particularly under conditions of low soil nitrogen. The presence of nitrogen in the spoil inhibits nodule nitrogen fixation. The tropical legume, soybean is often considered to be the most important grain legume in terms of its economic value. It is a vital source of vegetable protein for food and animal feed worldwide. Soybean nodules are classified as having determinate growth and structure. Symbiotic nitrogen assimilation in soybean nodules produces ureides, which are exported to the plant tissues. The ureides are converted into amino acids, which are used for the synthesis of a wide range of essential molecules, such as nucleotides and proteins.

The determinate nature of soybean nodules means that they lose meristematic activity soon after formation and, hence, nodule growth is based largely on cell expansion that results in the formation of the mature nodule. Soybean nodules are further classified into crown nodules and lateral nodules, depending on where they develop and are positioned on the root system. Crown nodules develop from the stem or tap root whereas lateral nodules develop from the lateral roots, as illustrated in Figure 1A. In general, nodules have a defined and limited functional lifespan and they tend to senesce early relative to other major plant organs, particularly the leaves. Nodule senescence is associated with a decrease in symbiotic nitrogen fixation, a degradation of leghemoglobin, and the breakdown of symbiosis (Figure 1B) [3]. The lifespan of nodules on fast growing annual legumes, such as soybean, is about 10–12 weeks from the point of initiation to senescence [4]. New nodules can be produced on the lateral roots as the crown nodule senesce leading to a heterogeneous population in terms of age, the oldest nodules on the soybean plant being the crown nodules that form on the main root.

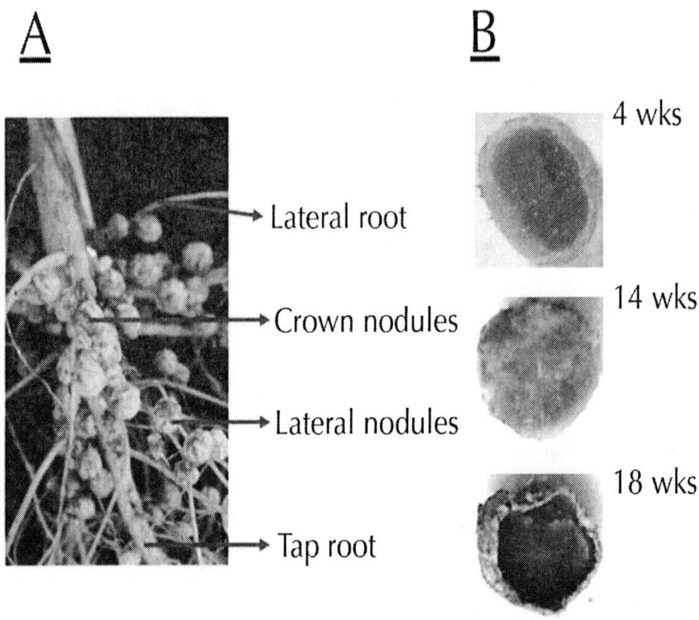

Figure 1: (A) Crown and lateral nodules on soybean (cultivar Prima 2000). Crown nodules develop early on the tap root (photo taken by E Cruywagen, University of Pretoria); (B) Sections of crown nodules at 4 weeks have a pink interior because of the presence of leghemoglobin associated with active nitrogen-fixation. At 14 weeks the interior becomes green as nodule senescence progresses from the center outwards. By 18 weeks the symbiosis has ended and the interior of the dying nodules becomes brown because of leghemoglobin degradation.

In contrast to the initiation of symbiosis and the first stages of nodule initiation, which have been intensively studied in recent years, very few studies have focused on the process of nodule senescence, which has been hardly been investigated in comparison. Although symptoms and progression of the processes of nodule death have been well described, the nature of the factors and mechanisms that control nodule senescence are largely unknown. Moreover, nodules can suffer premature senescence that is induced in response to abiotic stress factors, such as drought, and also by nitrate fertilization. The stress-induced changes that adversely affect nodule stability and performance are not well studied or understood. Premature nodule senescence shares some common characteristics with the developmental senescence process [5], starting at the center of the nodule and extending to the periphery [4]. Senescence in nodules as in other plant organs can be defined as a programmed degeneration process that is the final step of development leading to death [6]. The senescence process is characterized by the degradation of many metabolic and structural proteins, the nitrogen within the nodules being remobilized and translocated to the flowers to drive reproductive growth [7].

The induction of cysteine proteases during nodule senescence, which was first reported for peas, is a key process in the later stages of nodule development leading to proteolysis [8]. As in other organs, the induction of cysteine proteases is an excellent marker for nodule senescence [8]. The regulated turnover of proteins is important for cell functions and is an intrinsic part of the development process. It underpins cellular homeostasis and is required for the induction of programmed cell death (PCD) at the end of senescence [9]. All proteins are subject to turnover, involving degradation by the proteolytic systems of the cell including the proteosome and the suite of proteolytic enzymes. Proteins turn over at different rates according to their functions and sensitivity to degradation, which is enhanced during stress or when the nitrogen or carbon from the component amino acids is needed for other processes [10]. Cysteine proteases contain a catalytic sulfhydryl group at the active center involving a cysteine and a histidine residue. The majority of plant cysteine proteases belong to either the papain (C1) or legumain (caspases) (C13) families with legumains containing similar protein folds to caspases, which belong to the C14 family.

The activities of many proteases are regulated by protease inhibitors, which are generally small proteins that inactivate their protease substrates in a specific manner by forming reversible complexes (Figure 2) [11]. In the soybean nodule, the zone of high cysteine protease expression increases in size as the nodule develops and moves from the outer periphery toward the center of the organ [3]. Although the activities of cysteine proteases have been demonstrated during nodules senescence [12], little information is available about the genes that encode the cysteine protease forms that are expressed during early and late soybean nodule development. Moreover, to our knowledge, no detailed information exists in the literature concerning the expression of cysteine protease inhibitors (cystatins) during nodule development or in natural or stress-induced senescence.

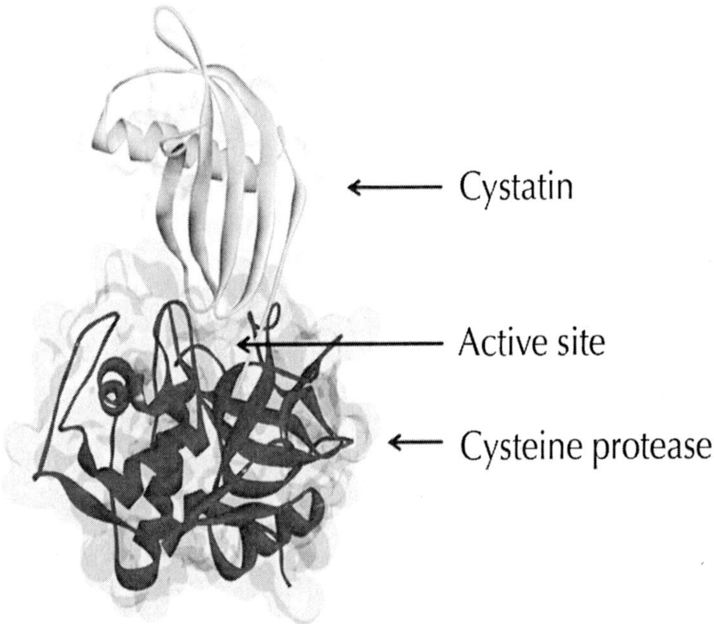

Figure 2: Cystatin binding to the active site of a cysteine protease.

Various biochemical and molecular tools have been used to characterize the cysteine proteases that are involved in nodule senescence. This includes the identification of activities and some analysis of gene expression profiles. In the following discussion, we summarize recent achievements in the characterization of the soybean nodule cysteine protease-cystatin system in relation to agronomic traits. As well as summarizing the relevant literature, we have included information from our own studies using different "omics" tools to characterize the cysteine proteases and cystatins that are specifically expressed at different stages of nodule development.

NODULE CYSTEINE PROTEASE-CYSTATIN SYSTEM

Cysteine Proteases

Plant proteases are responsible for a range of different processes within the context of protein metabolism. For example, cysteine proteases are central to the systems that dismantle organelles and degrade proteins for nutrient release from senescent tissues in order to translocate nutrients to actively growing tissues [13]. Cysteine proteases are therefore particularly abundant in senescing nodules. A subfamily of papain-like cysteine proteinases is regulated by senescence and stresses in soybean and these are considered to be particularly important in the orchestration of PCD [9,14,15]. Proteolytic enzyme activities have been described in both developing and senescent nodules [16]. Although cysteine proteases are expressed in nodules at the early stages of development, they are considered to be compartmentalized to prevent uncontrolled proteolysis that might otherwise impede symbiotic nitrogen fixation and ureide production [17,18]. Papain-like cysteine proteases have been isolated from legume nodules and germinating seeds [19]. More than 600 papain-like cysteine proteases can be found in the MEROPS database, which is an information resource for peptidases including proteases, proteinases and proteolytic enzymes and their inhibitors [20]. These C1A family proteases are classified according to their similarities to mammalian types of cathepsin L-, B-, H- and F-like in terms of their gene structures and phylogenetic relationships [21].

Papain-like cysteine proteases are synthesized in an inactive form in the endoplasmic reticulum and they are then transported to different organelles, particularly the vacuole and lysosomes [20]. Once in the target organelle, they undergo self-processing, or they are processed by enzymes, such as vacuolar processing enzymes (VPEs), to produce the active forms [22]. VPEs with caspase-like activities, also known as legumains, are aspartic cysteine proteases that are located in vacuoles or in the cell walls [23,24]. VPEs have different properties to other known cysteine proteases, even though they have cysteine and histidine residues at their active sites. For example, legumains are not sensitive to inhibition by E-64, a classic inhibitor of papain-like cysteine proteases. Moreover, they are active only at acidic pH values such as those found in vacuoles or in the cell wall/apoplast environment. VPE activities are responsible for the collapse of the vacuolar membrane releasing different proteases into the cell. They are also involved in other senescence processes and in the completion of PCD pathways [24]. In particular, the VPE forms are thought to be responsible for maturation of other proteases during senescence [25]. For example, a pre-pro- legumain region was found to cleave and activate a cathepsin B [26].

Nodule cysteine proteases fall into two classes, the nodule specific cysteine proteases and nodule enhanced proteases [27]. AsNODF32 is a nodule specific papain-like cysteine protease in the Chinese milk vetch that is activated during nodule senescence. This example of the first class shows highest transcript abundance in the senescent zone of the nodule, the level of transcripts decreasing gradually in the cell types distant from the senescence zone [28]. Temperate legumes, such as pea, alfalfa, and vetch, have indeterminate nodules that maintain zones of active apical meristematic activity and produce new cells throughout the nodule lifespan. Indeterminate nodules can, hence, house zones with very different developmental stages ranging from active meristematic zones, to zones with symbiosomes showing nitrogen fixation, to senescent zones where the breakdown of leghemoglobin is evident leading to a visible green tinge.

The lifespan of Chinese milk vetch nodules is extended when the expression of the AsNODF32 gene is silenced leading to enhanced nitrogen fixation [29]. The vacuolar cysteine protease, PsCyp15A, is expressed in the indeterminate nodules of pea, at a time when protease activity increases at the onset of senescence [15,30]. In experiments using anti-sense inhibition of CYP15A in *Medicago trunctula* nodules, the senescence process was delayed [31]. PsCYP15A exists in two forms, a pro-form (of approximately 38 kDa) and a mature form (of approximately 30 kDa), both forms being present in most tissues [30]. However, only the mature form was found in isolated symbiosomes containing nitrogen-fixing bacteroids. Immuno-labeling of nodule sections revealed that PsCYP15A was localized in large vacuolar bodies, cytoplasmic vesicles, and in the peri-symbiont space, suggesting that this protease is involved in protein turnover in the symbiosome compartment [30].

Cysteine proteases are expressed during the development of the determinate nodules found on soybean [3]. Soybean cysteine proteases are expressed around the periphery of the central infected zone, which is considered to be the developmental zone where nitrogen fixation is beginning to decline. Determinate nodules loose meristematic activity shortly after initiation, with subsequent cell expansion leading to the spherical shape of the mature nodule. We have identified several papain-like cysteine proteases in soybean crown nodules that have homology to CYP15A but little is known about the other cysteine proteases that are expressed in soybean nodules.

Cystatins

Protease inhibitors are involved in a wide range of plant defence mechanisms [20,21,29]. Cystatins or phytocystatins are inhibitors of papain-like cysteine proteases [11]. The entire cystatin gene family has been recently characterized from barley (*Hordeum vulgare*) and contains 13 non-redundant genes along with the cathepsin L-like target cysteine proteases [32]. Cystatins and cathepsin L-like cysteine proteases are found throughout the endoplasmic reticulum and the Golgi complex of the plant cell. Most of the cystatins contain signal peptides that direct them to the ER, where they are recognized as secretory proteins [33]. From the ER they are distributed to intracellular or extracellular locations. Most C1A proteases have specific localization sequences that deliver them to different cellular compartments or to secretion [34]. Within this system the cystatins can inactivate any cathespin L-like cysteine proteases that are present (Figure 2).

The transcriptional regulation of cystatin genes also plays an important role in the process of protein recycling during senescence processes and in stress conditions [11]. The expression of protease inhibitors is frequently decreased at times when proteases are up-regulated for example in senescent organs [35]. Cystatin mRNAs also increase in leaves exposed to abiotic stresses such as drought, salinity or low temperature [36]. Protease inhibitors are likely involved in the regulation of the nodulation process [37]. A Kunitz-type serine protease inhibitor, called SrPI1, is expressed at the early stages of nodulation particularly in the central zone of young nodules with lower expression levels mature nodules [29].

CYSTEINE PROTEASE AND CYSTATIN ACTIVITY IN SOYBEAN NODULES

Cysteine proteases and cystatins have been characterized in soybean crown nodules using both biochemical and transcript profiling approaches. Total nodule protease activities were measured in these studies using mildly denaturing gelatin-containing SDS-PAGE gel electrophoresis, in which proteins are not fully denatured and retain activity following separation on the gels [38]. Figure 3A illustrates the cysteine protease activities of soybean crown nodules as observed on gelatin-containing SDS-PAGE gels, where the unstained area has escaped gelatin degradation by proteases. The protease activities of soybean crown nodules of different ages measured in this way show almost identical patterns of total protease activity at least in samples up to an age of 10 weeks (Figure 3A; protease activity P1 and P2). In contrast, the older, senescent crown nodules showed an additional protease activity (P3) band with a higher protease activity in the P1 and P2 bands demonstrated by a more extensive clearance of the Coomassie blue-stained gel. When E-64, a potent inhibitor of papain-like cysteine proteases [39] was added to the gels less band clearance was observed in the senescent nodules. This observation would suggest that both E-64-sensitive proteases, most likely papain-like cysteine proteases, as well as cysteine proteases that are not sensitive to E-64 are expressed during soybean nodule senescence (Figure 3B).

The reverse zymogram technique can be used to detect the presence of cystatins in nodule extracts. This technique is based the ability of nodule cystatins to prevent the papain-catalyzed degradation of gelatin. After the electrophoretic separation of nodule proteins, the gelatin-containing gels were incubated with papain in order to degrade gelatin and nodule proteins. The gels are then stained with Coomassie blue and areas where nodule cystatins block papain, the gelatin remains stained [38]. Application of this technique revealed that cystatins are expressed in soybean nodules throughout development, except at the senescent stage (16–18 weeks old) where protease activities are high (Figure 3C). Down-regulation of transcription of cystatins and up-regulation and activation of cysteine proteases might have resulted in higher cysteine protease activity.

Fluorogenic substrates can be used to detect Phe-Arg-MCA cathepsin L-like, Z-Ala-Ala-Phe-MCA legumain-like and Z-Arg-MCA trypsin-like protease activities [39]. When these fluorogenic substrates are hydrolyzed by a protease, a bound -amino 4-methylcoumarin (MCA) derivative is released and this fluoresces in the unbound form. The hydrolysis of all these substrates was greatly increased in senescent nodules confirming that protease activities are increased at the later stages of development (Figure 4).

The major targets of proteases in root nodules are considered to be cytosolic proteins, such as leghemoglobin, which is the most abundant nodule protein, as part of symbiosome degradation [40]. The increases in protease activity, particularly in legumain activity, observed during nodule senescence occurred at the same time as leghemoglobin degradation (Figure 4D), as indicated by the green coloration of the nodules. The color change also indicates that leghemoglobin degrades from the center of the nodule, where the tissues change from pink to green/brown (Figure 1B). Nodule greening is due to biliverdin formation [41]. This color change has frequently been used as a visible marker of nodule senescence [17]. The primary role of leghemoglobin is the transport of oxygen, the high sensitivity of the nitrogenase enzyme in the *Rhizobium* bacteria to oxygen-dependent inactivation requiring tight control of tissue oxygen levels. The decrease in nitrogen fixation is associated with a decline in the ability to transport oxygen [42] as leghemoglobin degradation is initiated (Figure 5B).

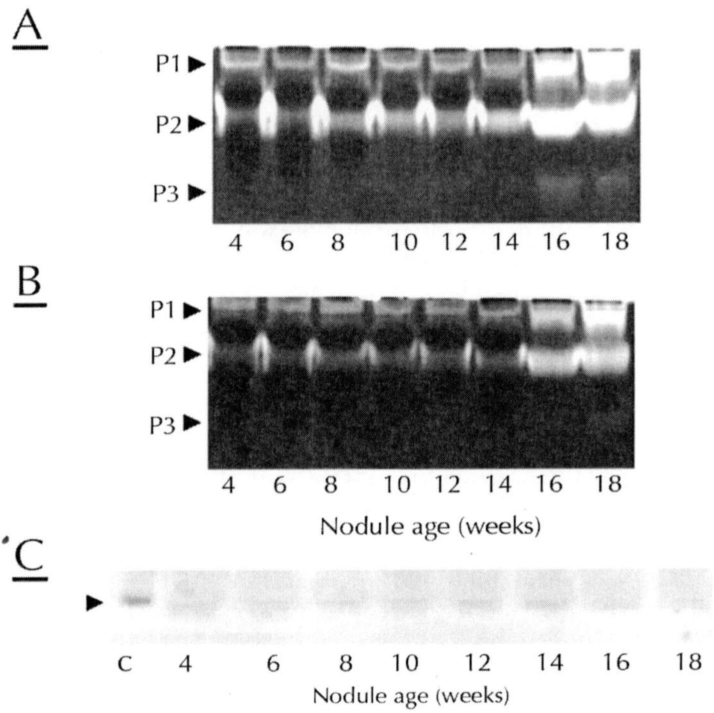

Figure 3: Total protease activities measured in soybean crown nodules (cultivar Prima 2000). Protease activities were visualized on a mildly denaturing gelatin-containing SDS-polyacrylamide gel over a period of 18 weeks without (A) and with the addition of E-64 (added to the sample before electrophoresis to a final concentration 10 μM). This detects the presence of both E-64 sensitive proteases (papain like cysteine proteases) as well as proteases from other classes not sensitive to E-64; (B) Reverse zymogram to detect cystatin activity against papain action (C); Different protease activities (P1, P2, and P3) are shown by clearance of stained gel. Gel electrophoresis was carried out with A Bio-Rad (UK) electrophoresis system on a 15% (w/v) sodium dodecyl sulfate polyacrylamide gel (SDS-PAGE) according to the method of Laemmli [43] containing 1% (w/v) of gelatin with 50 μg of protein per sample loaded. Protease re-naturing, gel treatment with papain (2mg/100mL of buffer, Sigma, USA) for reverse zymography for protease inhibitor detection and all gel staining procedures were carried out according to the method described by Michaud [38]. P represents detected protease activity and C oryzacystatin-I (1 μg) used as a control.

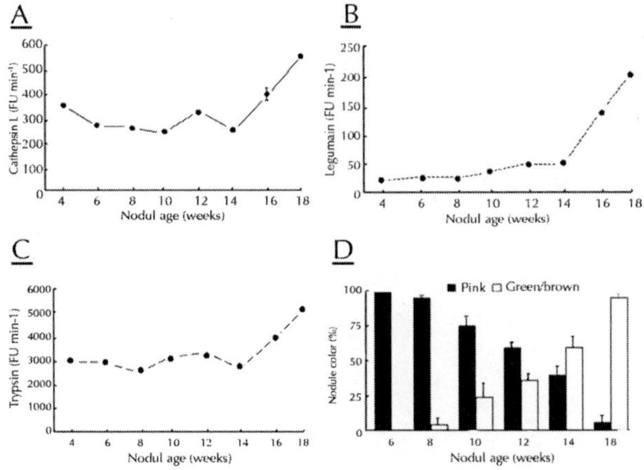

Figure 4: Determination of cathepsin Ll-like (A), legumain-like (B), trypsin-like (C) protease activities in soybean crown nodules (cultivar Prima 2000) and crown nodules color (D) over a period of 18 weeks. Cathepsin L-like activity with substrate Z-Phe-Arg-MCA was measured in 50 mM sodium phosphate buffer, pH 6.0, containing 5 mM l-cysteine, legumain-like cysteine protease activity with substrate Z-Ala-Ala-Asn-MCA measured in buffer containing 39.5 mM citric acid, 121 mM Na_2HPO_4, 1 mM DTT, 1 mM EDTA, and 0.01% CHAPS at pH 7.0 and trypsin-like activity with substrate Z-Arg-MCA in 100 mM Tris-HCl buffer at pH 8.0. Hydrolysis of the substrate was monitored using a hydrolysis progress curve [44]. Hydrolysis was monitored at 25 °C with a spectro-fluorometer (BMG FluoStar Galaxy) with excitation and emission at 340±10 nm and 450±10 nm, respectively. The reaction was measured over a period of 20 min. Reaction rates represented by the slope of the curve were measured as Fluorescence Units (FU/min). All reactions were performed in triplicate. For nodule color pink color indicates active leghemoglobin, greenish/brown color inactive leghemoglobin. Color values shown represent the mean of four individual experiments ± SE and protease activity data represent the mean ± SE of three individual activity measurements with 2 μg of nodule protein extract added to each assay.

Increased protease activities in senescent nodules are associated with both a decrease in nodule protein content and in nitrogenase activity (Figure 5C,D). The nitrogen-fixing capacity of nodules peaks early in the soybean nodule development (Figure 5D). Despite some drawbacks, the acetylene reduction assay method for the detection of nitrogenase activity is widely used because of its high sensitivity and simplicity. Nodule weight and number are also frequently used as markers of nitrogen fixation capacity [45]. Although the fresh weight of soybean nodules increased with time, the number of crown nodules was fixed early in development was constant thereafter (Figure 5A,B). However, the crown nodules increase in weight during development with the larger nodules fixing more nitrogen than the smaller nodules [46].

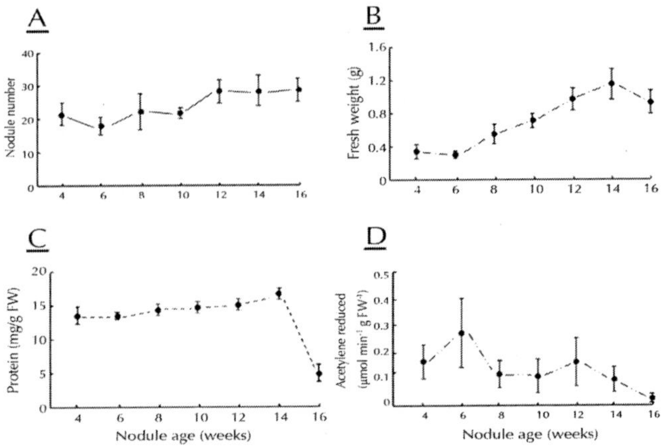

Figure 5: Production of nodule number (A), biomass (B), total protein content (C) and nitrogenase activity of soybean crown nodules (D) (cultivar Prima 2000) measured as μmol acetylene reducedmin^{-1} g FW^{-1} over a period of 16 weeks. For nodule formation soybean seeds (Glycine max L. Merr.; cultivar Prima 2000) were inoculated with *Bradyrhizobium japonicum* (strain WB 74-1) powder at 109 CFU g^{-1} (Soygro bio-fertilizer Limited, South Africa) and grown in pots [17.5 cm × 20 cm diameter (top) and 13.1 cm] in fine-grade vermiculite (Mandoval PC, South Africa) in an environmentally controlled green-house with a 13 h photoperiod at 25 °C/16 °C day/night temperature, 600 mmol m^{-2} s^{-1} photosynthetically active radiations (PAR) and 60% relative humidity and watered twice a week with de-ionized water and three-times-a-week with a nitrogen-free Hoagland solution. Protein content of nodule extracts was determined using a commercial protein determination kit (Bio-Rad, UK). Nitrogenase activity of nodules was determined with a gas chromatograph (Varian 3900; Varian Inc., USA) according to the method described by Turner and Gibson [47]. Data represent the mean ± SD of fresh weight and nodule number derived from three individual plants. Data represent the mean ± SD of activity derived from nodules of three individual plants.

CHARACTERIZATION OF THE CYSTEINE PROTEASE AND CYSTATIN TRANSCRIPTOME

Transcriptome profiling has been used to characterise the genes involved in the senescence of indeterminate nodules of *Medicago truncatula* [48]. Four cysteine proteases genes were highly expressed in nodule senescence. These genes were highly homologous to one of the most frequently used markers of leaf senescence, the cysteine protease encoded by the gene called SENESCENCE ASSOCIATED GENE 12 (*SAG12*) [19]. An early transcriptome profiling comparison was made on the nodules of two soybean cultivars that differ in sensitivity to dark chilling [46]. Microarray analysis was performed on nodules at seven weeks and 11 weeks, points where nitrogenase activity was maximal and where it had decreased to 50% of maximal rates [49]. The cultivars used in this study were Highveld Top, which is nominally chilling resistant and PAN 809, which is nominally chilling sensitive. While only 10% of the mRNAs that showed age-dependent changes in abundance could be annotated, 220 showed a common developmental pattern in both soybean cultivars with 180 genes up-regulated and 40 genes down-regulated in the nodules at the early senescent stage [49]. About 17% of up-regulated gene sequences belonged to transcription factor families. The dataset of senescence up-regulated genes also included those encoding enzymes involved in metabolism, such as cell wall modification and in general metabolic regulation, as well as stress responses [49]. Seven of the identified gene sequences were involved in protein degradation and these were up-regulated in the nodules at the early senescent stage. These included a cysteine protease, two peptidases and four putative trypsin (serine protease) inhibitors. The cysteine protease (Gma.8481.1.S1_at) that was up-regulated in the nodules at the early senescent stage belongs to the group of vacuolar processing enzymes (VPEs). However, neither papain-like cysteine proteases nor cystatins were changed in expression in the soybean nodules of either cultivar at the early senescent stage.

The soybean genome data base [50,51,52,53,54,55] has facilitated rapid identification of soybean cysteine proteases and their inhibitors. The sequences in the databases include 52 putative papain-like cysteine protease gene sequences, eight legumain-like cysteine proteases and 18 putative gene sequences with homology to the rice cystatin-I. This information has allowed a more in-depth analysis of the changes in transcripts encoding the different proteins of the cysteine protease–cystatin system in soybean crown nodules during development. RNA sequencing (RNA-Seq) of crown nodules (authors own unpublished data) revealed that papain-like and legumain-like proteases (VPEs) as well as cystatins are transcribed in crown nodule. These genes were amplified and sequenced so that identity could be confirmed by alignment with sequences from other species using database comparisons using Blast N and ClustalW. Semi-quantitative real-time PCR analysis was performed on crown nodules harvested at different stages of development (four, eight, and 14 weeks) to detect amplified products that had 100% homology to accessions used as templates. Only two of the sequences that encode putative papain-like cysteine proteases produced visible amplification products after 26 PCR cycles (Figure 6A, Table 1), the other papain-like cysteine proteases tested requiring up to 40 cycles for production of a visible amplification product. These data suggest that the two papain-like cysteine proteases (Glyma04g04400 and Glyma14g40670) detected after 26 PCR cycles are the most abundant papain-like cysteine proteases in nodules. However, the abundance of these transcripts was not increased during senescence. A much greater amount of amplification products was found for the sequences encoding legumain-like cysteine proteases (Glyma17g14680 and Glyma14g10620). In contrast to young nodules, which yielded little or no detectable amplification, high amplification of these cDNAs was observed when 26 PCR cycles were applied to extracts from the 14 week-old nodules (Figure 6B), suggesting that these transcripts were greatly increased in the senescent nodules. This result confirms a previous report showing increased expression of the legumain-like cysteine protease GmCysP1 in senescent nodules [12]. A database search revealed that GMCysP1 is identical to Glyma17g14680. The expression of legumain-like cysteine proteases is likely to be required in order to release and activate the inactive cysteine proteases that have been stored in the vacuoles and other compartments of the nodules. Cystatin are also expressed during nodule development. In this case amplification products of only two cystatins, Glyma14g04250 and Glyma05g28250, were detectable after 30 PCR cycles (Figure 6C), with some increase in the levels of Glyma14g04250 transcripts in senescent nodules.

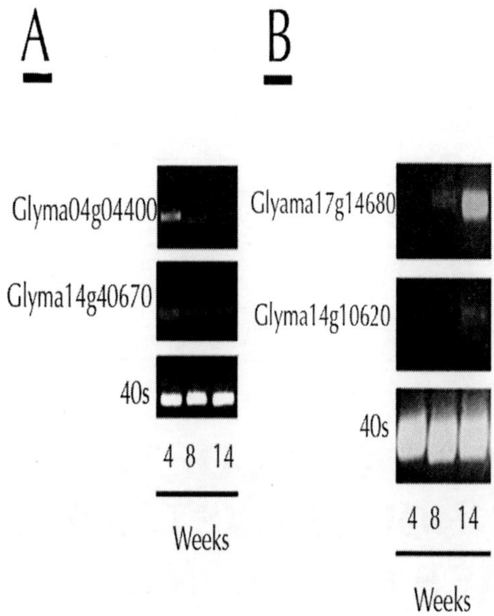

Figure 6: Transcription profiles of five papain-like cysteine proteases in soybean tissues of plants (cultivar Prima 2000) of different age (4, 8, and 14 weeks) determined by semi-quantitative RT-PCR. Amplification of the 40S gene sequence was used as a control. Semi-quantitative real-time polymerase chain reactions (semi-qRTPCR) were performed using 200 ng cDNA. For the reaction, 1 μL of the synthesized cDNA was used in a PCR reaction containing 2 μL of 10× PCR buffer, 1.6 μL of 25 mM $MgCl_2$, 2 μL of 2 mM deoxyribonucleotide triphosphates (dNTPs), 0.4 μL of primers (10 μM), 0.2 μL of five units *Taq* polymerase. DNA amplification was done in a thermal cycler (Bio-Rad, USA) with denaturing DNA at 94 °C for 5 min followed by 25 to 46 cycles depending on the amplification product consisting of 94 °C for 30 s, 55 °C for 30 s for DNA annealing and 72 °C for 1 min for DNA extension. Final DNA elongation at 72 °C for 10 min. PCR samples were run on a 2% (w/v) agarose gel and DNA was stained with GelRed™ (Biotum, USA) added

to 6X Loading Dye (Fermentas, Canada).

Table 1: Forward and reverse primers used for amplification of different papain- and legumain-like cysteine proteases and cystatins with size of amplicons produced and cycles required to obtain an amplification product

Accession number	Forward Primer	Reverse Primer	Amplicon (bp)	Cycles
Papain-like Glyma04g04400	GATCTTTAATGGCCACGATCCTCAT	CAGCACCTTGAAAGGGGTAATCCT	678	26
Glyma14g40670	ATATGGAGCGTGTGACTCGG	GTAATATCCATTCTCTCCCCAGCTC	430	26
Legumain-like Glyma17g14680	CTACGGAAACTACAGGCATC	GTTCTCCGTCGTCACATTAT	217	26
Glyma14g10620 *Cystatins*	GGTCGTGGATGTTGCTGAGG	ATCTGCTTGATGCCTGTAGTTTCC	191	26
Glyma14g04250	CACCGAAAGAGGATTAACAG	GGAGTTTGTGAGGGTGATTA	180	30
Glyma05g28250	GGATTAAAGCATACTAAACCA	GAATATTCGAATCCGTTGT	161	30

CHARACTERIZATION OF THE SOYBEAN NODULE CYSTEINE PROTEASE PROTEOME

When combing with measurements of the hydrolysis of protease substrates, the transcriptomics data provide a wealth of data allowing the identification of the protease genes and their substrates. However, protease activity often not depends on translation of transcripts into proteins but also post-translational modifications of proteins. The DCG-04 tagging technique, used together with visualization of tagged proteases by a streptavidin-based detection system, can assist the accurate identification of proteases and their substrates [56]. DCG-04 is an analog of the papain-like cysteine protease inhibitor E-64 that binds to these enzymes in an irreversible manner. DCG-04 carries a biotin residue allowing the tagging of substrate cysteine proteases by detection of the biotin carrying tagged proteins with peroxidase-labeled streptavidin, as illustrated in Figure 7A. Like E64, DCG-04 reacts with active cysteine proteases in a mechanism-based manner. Enrichment of the papain-like cysteine protease proteome using this tagging technique, first by incubation with DCG-04, followed by the separation of tagged proteases by SDS-polyacrylamide gel electrophoresis and blotting the DCG-04-tagged proteins onto membranes and identification using a chemiluminescene substrate, revealed DCG-04 tagged protein bands with a molecular mass of between 25 and 35 kDa that were expressed in crown nodule during development, as illustrated in Figure 7B. Detection in a one-dimensional SDS gel system revealed a strongly tagged band in extracts from four-week-old nodules. This band could not be related to an increase in total protease activity (Figure 3A) or to an increase in cathepsin-L like activity (Figure 4A). However, tagged proteins with weaker detection signals were more intense in the senescent 16 and 18 week-old nodules.

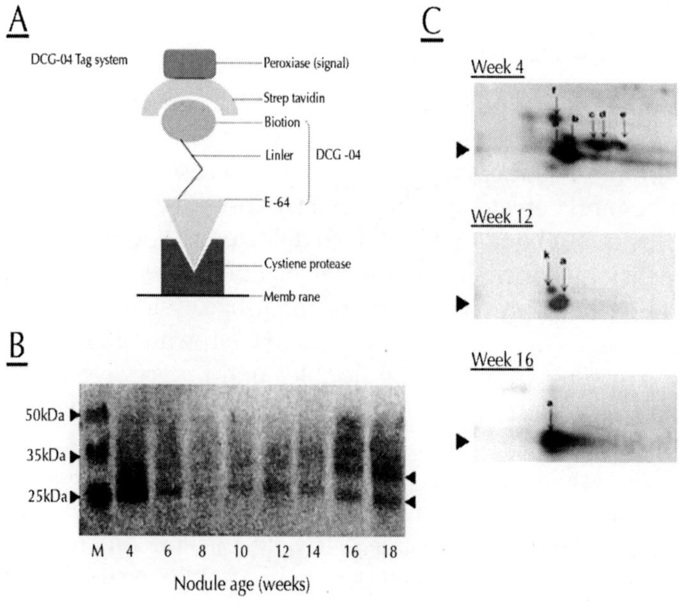

Figure 7: (A) Streptavidin-based detection system for cysteine proteases where peroxidase-labeled streptavidin binds to biotin linked with DGC-04 which is an E-64 analogue. DGC-04 binds to the active site of the cysteine proteases allowing detection of DCG-04-bound cysteine proteases via peroxidase activity measurement; (B) Detection of DCG-04-labeled cysteine proteases produced in 4 to 18 weeks old soybean crown nodules (cultivar Prima 2000). Arrows on the right indicate position of detected cysteine protease bands. M represents different size protein markers. For labelling and detection of DCG-

04 bound cysteine proteases with a streptavidin-based detection system the method described by Martinez *et al.* [57] was applied with five plant samples pooled; (C) Two-dimensional PAGE for detection of DCG-04-labeled cysteine proteases produced in 4 to 16 weeks old crown nodules. For first dimension, acetone precipitated nodule proteins (40 µg) were first added to 130 µL of the sample rehydration solution (8 M urea, 2 M thiourea, 2% CHAPS as a detergent, 50 mM DTT as a reducing agent, 0.02% ampholyte solution, and 5% bromophenol blue as a dye) in order to denature and solubilise the sample proteins for isoelectric focusing using ZOOM IPG Runner system according to the supplier's instructions (Life Technologies, UK). Second dimension was performed as SDS-PAGE according to Laemmli [43] using a 15% polyacrylamide gel and Bio-Rad mini-protean III apparatus Bio-Rad, UK). Protein spots were blotted onto a nitrocellulose membrane and the membrane was developed using the Thermoscientific enhanced chemiluminescent kit according to manufacturer's instructions (Thermo Fisher Scientific, USA) and visualized on a Syngene GeneSnap imaging system (Syngene, UK).

Two-dimensional polyacrylamide gel electrophoresis was also used to characterize the papain-like DCG-04 tagged cysteine proteases further. In these studies, tagged cysteine proteases were blotted onto membranes and detected by the streptavidin-peroxidase system. In this system, the major band that had been detected in young nodules using one-dimensional polyacrylamide gel electrophoresis, separated into several spots that represent different papain-like cysteine proteases with a molecular mass of about 30 kDa (Figure 7C). The number of detectable spots was greatest in the young nodules and progressively decreased as the nodules senesced such that only one major spot was detectable in the 16 week-old nodules. This analysis indicates that a number of papain-like cysteine proteases are present at the early stages of nodule development. Further analysis is required to determine the identities of these proteases and their substrates. To date, the enrichment of cystatins by using the streptavidin-tagged papain has not been successful.

NEXT GENERATION SEQUENCING A NEW "OMICS" TOOL

The availability of the soybean genome, in addition to recent advances in "omics" technologies, has greatly increased the applications of molecular tools to study gene function [58]. Advances in sequencing technologies have revitalized genome sequencing approaches that can be used to explore gene function, particularly in high-throughput system that are increasingly replacing traditional "gene-by-gene" approaches. Next generation sequencing tools, such as high-throughput RNA sequencing and the establishment of RNA-Seq atlases, will also dramatically improve transcriptome analysis and allow specific and high magnitude determination of gene transcription [59,60]. These technologies will extend the range of information produced by current Affymetrix GeneChip technology atlases for gene expression, particularly by direct comparisons of transcriptomes in different tissues, or at different times of expression or under varying environmental condition. The soybean RNA-Seq atlas can already be explored at SoyBase [51], which holds data for high-resolution gene expression in tissues such as roots and nodules at different developmental stages [61]. Mining of such RNA-Seq atlases can result in the discovery of novel candidate genes that are potentially involved in plant development and in stress responses.

We have recently applied Illumina sequencing to establish a RNA-Seq data base for soybean nodule development and to search for changes in the expression of cysteine proteases and cystatins. Our aim is to identify unique genes that are expressed during soybean nodule development and in stress-induced senescence, with a view that this information might be used effectively in the future to prevent senescence-associated decreases in nitrogen fixation and nodule protein degradation. In addition, large-scale RNA-Seq studies can be used together with DNA microsatellites (small sequence repeats or SSRs) to identify individual members of the cysteine protease family. Microsatellites could be associated with cysteine protease-dependent traits and therefore be useful as molecular markers in soybean breeding programs.

CONCLUSIONS

There is accumulating evidence that proteases, including cysteine proteases, are produced during legume nodule development. The data presented here have been used to illustrate that cysteine proteases are present at all stages of nodule development but that the composition of expressed forms is different in senescent nodules from that in young nodules. The expression and hence function of certain cysteine proteases is specific to nodule senescence allowing protein degradation and loss of nitrogen fixation. The data presented here indicate that legumain-like cysteine proteases, which are required to release and active papain-like cysteine proteases, are the major cysteine proteases that are expressed in senescent nodules, as illustrated in Figure 8, which provides a schematic overview of expression changes for the cysteine protease–cystatin system in during soybean nodule development.

Several papain-like cysteine proteases are expressed in young nitrogen fixing nodules but this expression does not appear to greatly alter the overall cysteine protease activity of the nodules. There is no evidence to date concerning cystatin involvement in soybean nodule development. We have, as yet, been unable to find any significant increases in the transcription of cystatins during nodule development. Cystatins fulfill important roles in biotic stress protection [11]. It is possible therefore to speculate that they may function in nodule development to prevent the papain-like cysteine protease-mediated proteolysis in nodules exposed to stress that would lead to stress-induced senescence. The finding that the ectopic over-expressing the rice cystatin-I in transgenic tobacco plants allowed a better recovery of photosynthesis after exposure to cold stress supports this view [62]. There is also evidence that some phytocystatins contain a carboxy-terminal extension with an amino acid motif (SNSL) similar to that involved in the inhibition of legumain-like proteins by human cystatins [63]. We have already identified phytocystatins in soybean nodules and we are currently testing the hypothesis that these cystatins might act as legumain inhibitors regulating legumain activity during nodule development. The above discussion has illustrated the fact that very little is currently known about cystatin function in nodules and that further research is vital to determine the role of the complete cysteine protease–cystatin system in nodule development under optimal and stress conditions.

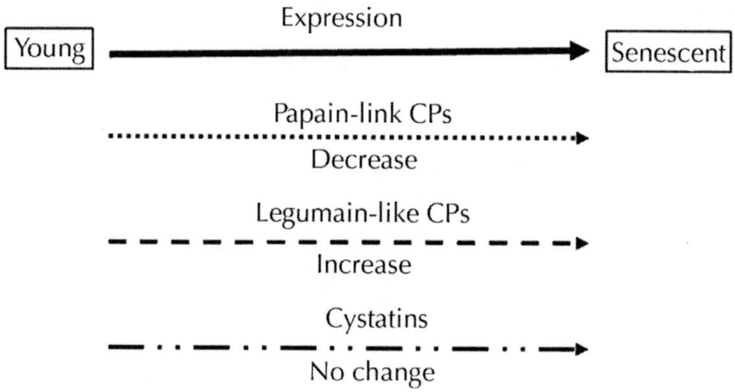

Figure 8: Schematic diagram of different types of cysteine proteases (CP) and cystatins expressed in soybean nodules at different development stages.

ACKNOWLEDGMENTS

This work was funded by FP7-PIRSES-GA-2008-230830 (LEGIM) and a Royal Society-NRF (Human and Institutional Capacity Development Programme) grant (NRF-2068793). Barend Juan Vorster acknowledges funding from the International Foundation of Science (IFS grant C/5151-1) and the NRF Thuthuka program and the Genomics Research Institute at the University of Pretoria. Stefan van Wyk and Matome Eugene Makgopa thank the NRF/DST and the Protein Research Foundation (Magdeleen du Plessis) in South Africa for bursaries. Marian Dorcas Quain thanks the Schlumberger Foundation Faculty for the Future Award for her fellowship and Urte Schlüter. the Deutsche Forschungsgemeinschaft and University of Pretoria for a post-doctoral fellowship. Our sincere thanks to Professor Matthew Bogyo (Department of Pathology, School of Medicine, Stanford University, USA) for providing the labeled DCG-04 probe.

REFERENCES

1. Lindemann, W.C.; Glover, C.R. Nitrogen Fixation by Legumes. Department of Agriculture Cooperating, New Mexico

State University, 1990. Available online: http://www.csun.edu/~hcbio027/biotechnology/lec10/lindemann.html (accessed on 12 July 2013).

2. Tilman, D.; Cassman, K.G.; Matson, P.A.; Naylor, R.; Polasky, S. Agricultural sustainability and intensive production practices. *Nature* 2002, *418*, 671–677.
3. Alesandrini, F.; Frendo, P.; Puppo, A.; Hérouart, D. Isolation of a molecular marker of soybean nodule senescence. *Plant Physiol. Biochem.* 2003, *41*, 727–732.
4. Puppo, A.; Groten, K.; Bastian, F.; Carzaniga, R.; Soussi, M.; Lucas, M.M.; de Felipe, M.R.; Harrison, J.; Vanacker, H.; Foyer, C.H. Legume nodule senescence: Roles for redox and hormone signalling in the orchestration of the natural aging process. *New Phytol.* 2005, *165*, 683–701.
5. Matamoros, M.A.; Baird, L.M.; Escuredo, P.R.; Dalton, D.A.; Minchin, F.R.; Iturbe-Ormaetxe, I.; Rubio, M.C.; Moran, J.F.; Gordon, A.J.; Becana, M. Stress-induced legume root nodule senescence. Physiological, biochemical, and structural alterations. *Plant Physiol.* 1999, *121*, 97–112.
6. Lim, P.O.; Woo, H.R.; Nam, H.G. Molecular genetics of leaf senescence in Arabidopsis. *Trends Sci.* 2003, *8*, 272–278.
7. Martinez, D.E.; Bartoli, C.G.; Grbic, V.; Guiamet, J.J. Vacuolar cysteine proteases of wheat (*Triticum aestivum* L.) are common to leaf senescence induced by different factors. *J. Exp. Bot.* 2007, *58*, 1099–1107.
8. Kardailsky, I.V.; Brewin, N.J. Expression of cysteine protease genes in pea nodule development and senescence. *Mol. Plant Microbe Inter.* 1996, *9*, 689–695.
9. Solomon, M.; Belenghi, B.; Delledonne, M.; Menachem, E.; Levine, A. The involvement of cysteine proteases and protease inhibitor genes in the regulation of programmed cell death in plants. *Plant Cell* 1999, *11*, 431–443.
10. Estelle, M. Proteases and cellular regulation in plants. *Curr. Opin. Plant Biol.* 2001, *4*, 254–260.
11. Benchabane, M.; Schlüter, U.; Vorster, J.; Goulet, M.C.; Michaud, D. Plant cystatins. *Biochimie* 2010, *92*, 1657–1666.

12. Oh, C.J.; Lee, H.; Kim, H.B.; An, C.S. Isolation and characterization of a root nodule-specific cysteine proteinase cDNA from soybean. *J. Plant Biol.* 2004, *47*, 216–220.

13. Beers, E.P.; Woffenden, B.J.; Zhao, C. Plant proteolytic enzymes: Possible roles during programmed cell death. *Plant Mol. Biol.* 2000, *44*, 399–415.

14. Esteban-García, B.; Garrido-Cárdenas, J.A.; Alonso, D.L.; García-Maroto, F. A distinct subfamily of papain-like cystein proteinases regulated by senescence and stresses in *Glycine max*. *J. Plant Physiol.* 2010, *167*, 1101–1108.

15. Lee, H.; Hur, C.G.; Oh, C.J.; Kim, H.B.; Pakr, S.Y.; An, C.S. Analysis of the root nodule-enhanced transcriptome in soybean. *Mol. Cells* 2004, *18*, 53–62.

16. Asp, T.; Bowra, S.; Borg, S.; Holm, P.B. Molecular cloning, functional expression in *Escherichia coli* and enzymatic characterisation of a cysteine protease from white clover *Trifolium repens*. *Biochim. Biophys. Acta Proteins & Proteomics* 2004, *1699*, 111–122.

17. Pfeiffer, N.E.; Torres, C.M.; Wagner, F.W. Proteolytic activity in soybean root nodules: Activity in host cell cytosol and bacteroids throughout physiological development and senescence. *Plant Physiol.* 1983, *71*, 797.

18. Pladys, D.; Vance, C.P. Proteolysis during development and senescence of effective and plant gene-controlled ineffective alfalfa nodules. *Plant Physiol.* 1993, *103*, 379–384.

19. Lohman, K.N.; Gan, S.; John, M.C.; Amasino, R.M. Molecular analysis of natural leaf senescence in *Arabidopsis thaliana*. *Physiol. Plant.* 1994, *92*, 322–328.

20. Martinez, M.; Cambra, I.; González-Melendi, P.; Santamaría, M.E.; Díaz, I. C1A cysteine-proteases and their inhibitors in plants. *Physiol. Plant.* 2012, *145*, 85–94.

21. Martinez, M.; Diaz, I. The origin and evolution of plant cystatins and their target cysteine proteinases indicate a complex functional relationship. *BMC Evol. Biol.* 2008, *8*, 198.

22. Wiederanders, B. Structure-function relationships in class CA1 cysteine peptidase propeptides. *Acta Biochim. Pol.* 2003, *50*, 691–713.

23. Müntz, K.; Shutov, A.D. Legumains and their functions in plants. *Trends Plant Sci.* 2002, *7*, 340–344.
24. Hara-Nishimura, I.; Hatsugai, N.; Nakaune, S.; Kuroyanagi, M.; Nishimura, M. Vacuolar processing enzyme: An executor of plant cell death. *Curr. Opin. Plant Biol.* 2005, *8*, 404–408.
25. Roberts, I.N.; Caputo, C.; Criado, M.V.; Funk, C. Senescence-Associated proteases in plants. *Physiol. Plant.* 2012, *145*, 130–139.
26. Sajid, M.; McKerrow, J.H.; Hansell, E.; Mathieu, M.A.; Lucas, K.D.; Hsieh, I.; Greenbaum, D.; Bogyo, M.; Salter, J.P.; Lim, K.C. Functional expression and characterization of *Schistosoma mansoni* cathepsin B and its trans-activation by an endogenous asparaginyl endopeptidase. *Mol. Biochem. Parasit.* 2003, *131*, 65–75.
27. Li, Y.; Zhou, L.; Li, Y.; Chen, D.; Tan, X.; Lei, L.; Zhou, J. A nodule-specific plant cysteine proteinase, AsNODF32, is involved in nodule senescence and nitrogen fixation activity of the green manure legume *Astragalus sinicus*. *New Phytol.* 2008, *180*, 185–192.
28. Naito, Y.; Fujie, M.; Usami, S.; Murooka, Y.; Yamada, T. The involvement of a cysteine proteinase in the nodule development in Chinese milk vetch infected with *Mesorhizobium huakuii* subsp. rengei. *Plant Physiol.* 2000, *124*, 1087–1096.
29. Lievens, S.; Goormachtig, S.; Holsters, M. Nodule-Enhanced protease inhibitor gene: Emerging patterns of gene expression in nodule development on *Sesbania rostrata*. *J. Exp. Bot.* 2004, *55*, 89–97.
30. Vincent, J.L.; Brewin, N.J. Immunolocalization of a cysteine protease in vacuoles, vesicles, and symbiosomes of pea nodule cells. *Plant Physiol.* 2000, *123*, 521–530.
31. Sheokand, S.; Dahiya, P.; Vincent, J.; Brewin, N. Modified expression of cysteine protease affects seed germination, vegetative growth and nodule development in transgenic lines of *Medicago truncatula*. *Plant Sci.* 2005, *169*, 966–975.
32. Martinez, M.; Cambra, I.; Carrillo, L.; Diaz-Mendoza, M.; Diaz, I. Characterization of the entire cystatin gene family in barley and their target cathepsin l-like cysteine-proteases, partners in

the hordein mobilization during seed germination. *Plant Physiol.* 2009, *151*, 1531–1545.

33. Tian, L.; Zhang, L.; Zhang, J.; Song, Y.; Guo, Y. Differential proteomic analysis of soluble extracellular proteins reveals the cysteine protease and cystatin involved in suspension-cultured cell proliferation in rice. *Biochim. Biophys. Acta* 2009, *1794*, 459–467.
34. Grudkowska, M.; Zagdanska, B. Multifunctional role of plant cysteine proteinases. *Acta Biochim. Pol.* 2004, *51*, 609–624.
35. Sugawara, H.; Shibuya, K.; Yoshioka, T.; Hashiba, T.; Satoh, S. Is a cysteine proteinase inhibitor involved in the regulation of petal wilting in senescing carnation (*Dianthus caryophyllus* L.) flowers? *J. Exp. Bot.* 2002, *53*, 407–413.
36. Coupe, S.A.; Sinclair, B.K.; Watson, L.M.; Heyes, J.A.; Eason, J.R. Identification of dehydration-responsive cysteine proteases during post-harvest senescence of broccoli florets. *J. Exp. Bot.* 2003, *54*, 1045–1056.
37. Hong, J.K.; Hwang, J.E.; Lim, C.J.; Yang, K.; Jin, Z.L.; Kim, C.Y.; Koo, J.C.; Chung, W.S.; Lee, K.O.; Lee, S.Y. Over-expression of Chinese cabbage phytocystatin 1 retards seed germination in Arabidopsis. *Plant Sci.* 2007, *172*, 556–563.
38. Michaud, D. Gel electrophoresis of proteolytic enzymes. *Anal. Chim. Acta* 1998, *372*, 173–185.
39. Barrett, A.J.; Kembhavi, A.A.; Brown, M.A.; Kirschke, H.; Knight, C.G.; Tamai, M.; Hanada, K. l-Trans-epoxysuccinyl-leucylamido(4-guanidino) butane (E-64) and its analogues as inhibitors of cysteine proteinases including cathepsins B, H and L. *Biochem. J.* 1982, *201*, 189–198.
40. Gunther, C.; Schlereth, A.; Udvardi, M.; Ott, T. Metabolism of reactive oxygen species is attenuated in leghemoglobin-deficient nodules of *Lotus japonicus*. *Mol. Plant Microbe Inter.* 2007, *20*, 1596–1603.
41. Swaraj, K.; Bishnoi, N. Physiological and biochemical basis of nodule senescence in legumes: A review. *Plant Physiol. Biochem.* 1996, *23*, 105–116.
42. Appleby, C.A. Leghemoglobin and *Rhizobium* respiration. *Ann. Rev. Plant Physiol.* 1984, *33*, 443–478.

43. Laemmli, U.K. Cleavage of structural proteins during the assembly of the head of bacteriophage T4. *Nature* 1970, *227*, 680–685.

44. Salvesen, G.; Nagase, H. Inhibition of proteolytic enzymes. In *Proteolytic Enzymes: A Practical Approach*; Beynon, R.J., Bond, J.S., Eds.; IRL Press at Oxford University Press: Oxford, UK, 1989; pp. 83–104.

45. Al-Karaki, G.N. Morphological and yield traits of wild legume (*Tetragonolobus palaestinus* Boiss) populations. *J. Agron. Crop Sci.* 2000, *184*, 267–270.

46. Ashraf, M.; Iram, A. Drought stress induced changes in some organic substances in nodules and other plant parts of two potential legumes differing in salt tolerance. *Flora* 2005, *200*, 535–546.

47. Turner, G.L.; Gibson, A.H. Measurement of nitrogen fixation by indirect means. In *Methods for Evaluating Biological Nitrogen Fixation*; Bergersen, F.J., Ed.; John Wiley and Sons: New York, NY, USA, 1980; pp. 111–138.

48. Van de Velde, W.; Guerra, J.C.P.; De Keyser, A.; De Rycke, R.; Rombauts, S.; Maunoury, N.; Mergaert, P.; Kondorosi, E.; Holsters, M.; Goormachtig, S. Aging in legume symbiosis. A molecular view on nodule senescence in *Medicago truncatula. Plant Physiol.* 2006, *141*, 711–720.

49. Fenta, B.A.; Urte Schlüter, U.; Garcia, B.M.; DuPlessis, M.; Foyer, C.H.; Kunert, K.J. Identification and Application of Phenotypic and Molecular Markers for Abiotic Stress Tolerance in Soybean. In *Soybean—Genetics and Novel Techniques for Yield Enhancement*; Krezhova, D., Ed.; InTech: Shanghai, China, 2011; pp. 181–200.

50. Phytozome, Version 9.1: Biomart. Joint Genome Institute, Center for Integrative Genomics and University of California Regents: USA, 2006–2013. Available online: http://www.phytozome.net/biomart (accessed on 12 July 2013).

51. SoyBase and the Soybean Breeder's Toolbox. USDA-ARS and IOWA State University, 2010. Available online: http://soybase.org (accessed on 12 July 2013).

52. Grant, D.; Nelson, R.T.; Cannon, S.B.; Shoemaker, R.C. SoyBase, the USDA-ARS soybean genetics and genomics database. *Nucleic Acids Res.* 2010, *38*, 843–846.

53. LIS. Legume Information System, USDA-ARS and National Center for Genome Resources, 2013. Available online: http://comparative-legumes.org (accessed on 12 July 2013).
54. Legume Base. University of Miyazaki: Miyazaki, Japan, 2009. Available online: http://www.legumebase.brc.miyazaki-u.ac.jp (accessed on 12 July 2013).
55. Schmutz, J.; Cannon, S.B.; Schlueter, J.; Ma, J.; Mitros, T.; Nelson, W.; Hyten, D.L.; Song, Q.; Thelen, J.J.; Cheng, J.; *et al*. Genome sequence of the palaeopolyploid soybean. *Nature* 2010, *463*, 178–183.
56. Van der Hoorn, R.A.L.; Leeuwenburgh, M.A.; Bogyo, M.; Joosten, M.H.A.J.; Peck, S.C. Activity profiling of papain-like cysteine proteases in plants. *Plant Physiol.* 2004, *135*, 1170–1178.
57. Martínez, D.E.; Bartoli, C.G.; Grbic, V.; Guiamet, J.J. Vacuolar cysteine proteases of wheat (*Triticum aestivum* L.) are common to leaf senescence induced by different factors. *J. Exp. Bot.* 2007, *58*, 1099–1107.
58. Mochida, K.; Shinozaki, K. Advances in Omics and bioinformatics tools for systems analyses of plant functions. *Plant Cell Physiol.* 2011, *52*, 2017–2038.
59. Mardis, E. The impact of next-generation sequencing technology on genetics. *Trends Genet.* 2008, *24*, 133–141.
60. Tran, L.S.; Mochida, K. Functional genomics of soybean for improvement of productivity in adverse conditions. *Funct. Integr. Genomics* 2010, *10*, 447–462.
61. Severin, A.J.; Woody, J.L.; Bolon, Y.T.; Joseph, B.; Diers, B.W.; Farmer, A.D.; Muehlbauer, G.J.; Nelson, R.T.; Grant, D.; Specht, J.E.; *et al*. RNA-Seq Atlas of Glycine max: A guide to the soybean transcriptome. *BMC Plant Biol.* 2010, *10*, 160.
62. Van der Vyver, C.; Schneidereit, J.; Driscoll, S.; Turner, J.; Kunert, K.; Foyer, C.H. Oryzacystatin I expression in transformed tobacco produces a conditional growth phenotype and enhances chilling tolerance. *Plant Biotech. J.* 2003, *1*, 101–112.
63. Martinez, M.; Diaz-Mendoza, M.; Carrillo, L.; Diaz, I. Carboxy terminal extended phytocystatins are bifunctional inhibitors of papain and legumain cysteine proteinases. *FEBSLett.* 2007, *581*, 2914–2918.

Chapter 5

New Approaches for Crop Genetic Adaptation to the Abiotic Stresses Predicted with Climate Change

Robert Redden

Department of Primary Industries, Horsham Victoria, 3401, Australia

ABSTRACT

Extreme climatic variation is predicted with climate change this century. In many cropping regions, the crop environment will tend to be warmer with more irregular rainfall and spikes in stress levels will be more severe. The challenge is not only to raise agricultural production for an expanding population, but to achieve this under more adverse environmental conditions. It is now possible to systematically explore the genetic variation in historic local landraces by using GPS locators and world climate maps to describe the natural selection for local adaptation, and to identify candidate germplasm for tolerances to extreme stresses. The physiological and biochemical components of these expressions can be genomically investigated with candidate gene approaches and next generation sequencing. Wild relatives of crops have largely untapped genetic variation for abiotic and biotic stress tolerances, and could greatly expand the available domesticated gene pools to assist crops to survive in the predicted extremes of climate change, a survivalomics strategy. Genomic strategies can assist in the introgression of these valuable traits into the domesticated crop gene pools, where they can be better evaluated for crop improvement. The challenge is to increase agricultural productivity despite climate change. This calls for the integration of many disciplines from eco-geographical analyses of genetic resources to new advances in genomics, agronomy and farm management, underpinned by an understanding of how crop adaptation to climate is affected by genotype × environment interaction.

INTRODUCTION

Both genomics and phenomics will assist the investigation of genetic variation in both crop landrace germplasm and in crop wild relatives. Novel genetic variation will be needed to extend the range of tolerances to high temperature stresses and to severe droughts in the crop growing period, that are predicted to be more extreme with climate change this century [1,2]. In particular, the incidences of extreme conditions are expected to exceed those currently experienced in the temperate-tropical cropping regions [3]. This poses a threat to world food security, especially for production of the annual staple food crops of the world; wheat, rice, maize, sorghum and potato. A range of other crops from grain legumes to clonally propagated tubers, are also expected to suffer yield declines, especially when extreme conditions occur in the reproductive period of grain crops or in the tuberisation period for root crops [4]. Anthesis and pollination in many crops are adversely affected by temperatures above 35 °C, and the critical temperature may be as low as 32 °C [5]. High night temperatures above 32 °C can impede gametogenesis and photosynthesis in rice [6]. In potatoes the initiation of tubers requires temperatures below 18 °C.

Severe moisture stress in the pre-flowering and the flowering periods reduces formation of florets as well as affecting fertilisation and seed set [7]. Grain size is also reduced by terminal drought in the ripening period. Where rainfall is lower due to climate change, the reduced cloud cover may lead to increased frequency of frost in the reproductive period with further loss of yield potential.

The rise in global world temperature is now expected to exceed 2 °C by around 2050 in the worst case scenario, and certainly before 2100. In the temperate—tropical latitudes this will result in shorter crop growth periods and reduced yield potential [8]. There are some positive aspects to the predicted climate changes. Cropping will become more feasible at high altitudes, and at high latitudes, with longer periods above freezing. There will also be a fertilisation effect for more efficient photosynthesis due to higher global levels of CO_2 from increased industrial, urban and agricultural emissions. However this productivity response is very unlikely to compensate for severe losses from the predicted spikes in temperature and moisture stresses and overall more unfavourable crop growth environments [2,3].

World population growth is on schedule to exceed 9 billion by 2050, as well as being increasingly urbanised, and competing with agriculture for water and land resources. Even as major crop production environments become less favourable for crops, over 70% more production will be required to provide adequate food, mainly on the same area of arable land [8,9]. But to achieve this there is an urgent need to identify novel genetic variation, for crops to tolerate the expected levels of abiotic stresses beyond those experienced under the natural selection of domesticated crops dating back 11,000–13,000 years [2].

Integrated genetic (long term) and agronomic (short and long term) innovations will be important to raise world food production [10,11]. Many complementary strategies in conservation agriculture, weed control and chemical fallowing, innovative irrigation with the use of raised seed beds and of sub-surface drip systems, tie-bunding across cultivation ridges and inter-crop relay planting for more intensive land use, are some of the options that could provide tailored whole farm management specific to local environments [10]. Switching to more stress tolerant crops may be an option for particular crop regions, for example pearl millet is more drought tolerant than sorghum [12], triticale and rye are more drought tolerant than wheat with different respective strategies for maintaining photosynthesis, or for recovering from drought [13], and short season hybrid pigeonpea is preferred over maize in parts of east Africa [14]. A complementary option is to seek genetic variation for tolerance of abiotic stresses in both the domesticated crop genepool and that of its wild relatives, for each of the important plant food species. This paper explores the opportunities available in the gene pools of crops and their wild relatives to use new genomic tools for the next phase of plant breeding, and selection of crops with increased yield potential for future target environments of greater variability and extremes of temperature and moisture stresses than exist today.

ABIOTIC STRESS TOLERANCE IN THE DOMESTICATED GENEPOOLS OF CROPS

Landrace populations or traditional local varieties with a diversity of types and polymorphic genetic variation have been selected through natural and manual selection for population complexes with specific adaptation to the range of seasonal crop environments in each locality [15]. Between 1976 and 2003, in response to shorter and less reliable growing seasons with less rainfall in Niger (part of the west African Sahel), natural and human selection in landraces of pearl millet has increased earliness, reduced plant height and shortened reproductive spikes [16]. Landraces need to be productive and reliable across the yearly variation in growing conditions especially if rainfed, or they will be replaced by other varieties or alternate crops, especially for low input sustainable agriculture on family operated small farms [17,18]. Thus landraces through their distribution across the myriad of local environments worldwide reflect natural selection for changes in allele frequencies at loci controlling fitness traits for local adaptation, and have much wider genetic diversity than the varieties generated by modern plant breeding [19,20]. In the last century scientifically managed plant breeding programs have enabled gains in productivity in many crops, through using a very high selection intensity for a small number of key traits (increased harvest index, improved plant architecture), very large breeding populations, the pyramiding of complementary pest and disease resistances, and high crop input responsiveness, for various target environments [6,21].

Accurate geographic positioning of the locations where both wild and landrace germplasm is collected is now possible with GPS data on longitude, latitude and elevation. World climate maps are freely available on the internet, with a local resolution down to 1 square km [22], providing 25 years of historic climatic data for a given location. This climatic data corresponds to the farm where a landrace population was collected [23]. The coefficient of variation and the means for key plant growth variables can be tabulated for over 25 years and separated between the vegetative and reproductive growth phases, for maximum and minimum temperatures, pre-season and in-crop rainfall, relative humidity, sunshine hours, day-length, frequency of frost and also the frequencies of extreme stress spikes [24]. This provides a climatic record of the location where a population mixture of locally adapted genotypes comprising the landrace survived, in part as a result of natural selection for tolerance of abiotic stresses [15]. Such selection would be affected by; the available genetic variation within a landrace; the balance among genotypes variously selected over different seasons; land race effective population size (may be shared by a village); occasional introgression over the long term from newly arrived introductions of other landraces [25]; accumulated mutations over hundreds of generations; and selection for survival under the combined effects of abiotic and of biotic stresses plus manual selection [20]. Analysis of the combined site data for a region/country/world identifies the particular collection sites with a history of extreme abiotic stress, thereby prioritising the corresponding landraces for potential verification of this stress tolerance [23,26]. With this approach Li Ling *et al.* [24] short listed 61 pea accessions from 629 landraces in China, which survived under extremes of high temperature, drought, or frost in the reproductive growth phase. Both domestic and wild crop germplasm from locations with these stresses can be selected as candidates for potential tolerance to that stress, as a possible result of long term natural selection for survival either as a crop landrace population or as a wild relative.

Not all landraces from localities with a history of extreme stress are necessarily stress tolerant. Ehlers and Hall [27] screened 48 land races of cowpea (*Vigna unguicalata*) from the West African Sahel region to identify 2 with tolerance of high temperatures (up to 42 °C) during the reproductive phase. Although many had an early maturity escape mechanism, a small number exhibited high temperature tolerance with successful gametogenesis, fertilisation and seed set at temperatures above 40 °C in a desert environment. Such high temperature tolerance was also found in land races of vegetable cowpea (*Vigna sesquipedalis*) from northern India. These high temperature tolerances were verified under controlled growth cabinet conditions [28]. The use of geographic and agro-ecological information can assist in focussing the search for genes for stress tolerances from large genetic resource collections to just those landraces prioritised by respective collection environments with a history of extreme abiotic stresses. This can be augmented by appropriate sampling strategies [23]. The use of geographic mapping tools enables this search to be both more feasible and more practical.

NOVEL GENETIC STRESS TOLERANCE FROM WILD RELATIVES

A phenomenon known as a domestication bottleneck [29], or founder effect [30], took place with the domestication of crops from wild relatives [31]. Prior to the advent of agriculture food grains were collected direct from plants unless dispersed, or from the surrounding ground. Annual plants survived under natural selection through having efficient and effective dispersal mechanisms. These included; various forms of shattering or dis-articulation of seed from the flowering stalk such as seed pods twisting open to throw seed a short distance; awns to aid burial of seed; spines on seed coats to attach to passing animals; production of a many very small seed; or very hard/thick seed coats, which resisted digestion to enable seed dispersal through animal excreta, or delayed germination over a number of growth seasons [32]. In each environment, different plant species competed for survival with various modes of regeneration. The imperative was species survival with sufficient stored energy in the seed to enable germination of the embryo, and not on supplying a surplus of stored carbohydrates and proteins in seed to provide food and feed for humans and animals.

Selection for domestication was crop specific, but in grain crops typically required a shift in plant morphology and physiology; to retention of seed on plants for ease of harvest, to larger seed size so that agriculture became a more attractive food source than hunter-gathering, to a reduction in hard seed coats to allow reliable germination of annual crops, to improved nutrition and digestion, and for investment of more plant photosynthate in the seed rather than in vegetative growth [33]. Since these domestication traits were largely absent in the wild relatives there was a dependence on rare mutations to obtain the necessary major genes and quantitative variation for these traits. Such mutations needed to be observed and selected for, from natural recombinations of these different mutations, to enable successful both successful cropping and adaptation to local ecosystems [20,32]. The low level of cross-pollination in the self-pollinated crops (wheat, barley, rice, and the majority of grain legumes) largely isolated these crops from their wild relatives [34]. For cross pollinated crops barriers to genetic exchange with their wild progenitors through differences in flowering time, inhibitors of cross compatibility, or differences in ecological distribution; tend to separate the respective gene pools [35]. Thus the resultant genetic variation that was transferred to crops from wild progenitors was very restricted by these rare events, a domestication bottleneck.

This bottleneck limited transfer of the genetic diversity for abiotic stress tolerances from the much wider genepools of the crop progenitors and of other wild relatives. The secondary and tertiary wild relatives of crops may be rich sources of desirable traits such as disease resistance [36], yield and agronomic traits [37], or of genes for regulatory and physiological stress tolerance mechanisms [38]. The much wider genetic diversity available in wild relatives offer the possibility of a breakthrough in the apparent yield ceilings in many crop breeding programs (or reduced rates of yield gains), as demonstrated with the pyramiding of yield promoting genes in domestic tomato to achieve a 50% yield gains in both irrigated and dry field conditions [39].

The wild progenitors often can still be found in the respective regions where a crop was domesticated, growing wild often in marginal habitats unsuited to agriculture with severe abiotic stresses, or sometimes even as a weed in cultivated crops [40], but there is a risk of erosion of some of this diversity with climate change [23]. Wild progenitors vary in their distribution, being restricted to a small region in Turkey for *Cicer reticulatum* the progenitor of chickpea [41], to a wide distribution of *Pisum sativum* subspp. *elatius* the wild progenitor of pea which preceded crop migration with the spread of agriculture, and to spread of the wild progenitor as a weed in the cultivated crop as for *Hordeum spontaneum* the wild progenitor of barley [42]. To different extents for each crop, an overlap in distribution with the progenitor plus other cross compatible wild relatives continued after domestication, and rare (in self-pollinating species) to occasional (cross pollinating species such as sorghum) introgression of both beneficial and detrimental genes from wild to cultivated has continued post domestication in many crops [43]. Now however, the gene pools of many domestic crops can be selectively broadened with manually directed genetic introgression from wild relatives for improved tolerances to abiotic stresses of drought, high temperature and frost especially in the reproductive phase [37,44]. These three stresses are likely to pose the most important challenges of climate change to agriculture.

Wild relatives survive on uncultivated ground in generally harsher environments. Wider genetic variation for tolerance of abiotic stresses, as a result of natural selection for survival with better seed set on the more tolerant plants, is more likely to be available in wild relatives than in the genepools of most domestic crops [32,40].

ECOLOGICAL SCREENING BY THE HABITATS FOR COLLECTION OF GERMPLASM

A systematic preliminary screening of both landraces (efficiently sampled as core or mini-core collections representing the species diversity (23]), and wild relatives, based on the growing season history of their respective habitats, provides a short list of selected germplasm for more intensive evaluation, in an efficient and strategic search for the required genetic tolerance of abiotic stress. Not all candidates for stress tolerance will have this characteristic, as some may have escape mechanisms such as very early flowering; indeterminacy; very low biomass; capacity to recover from short term stress spikes, or may represent the genetic remnants from a narrow genetic base as the species expanded to new ecological niches [45,46]. Such escape mechanisms may suffice in current crop environments but may be inadequate to manage the more intense stresses predicted with climate change [2].

The chances of locating sources of genetic tolerance to abiotic stresses are greatly improved by the screening the environments of sites where germplasm was collected [23], noting that land race populations have wider genetic variance than conventionally bred varieties, and that wild relatives will have yet wider genetic diversity in most cases [20,31].

This approach, "survivalomics", is useful to identify parental material for introgression of abiotic stress tolerance based on survival in extreme habitats. There may be extensive epistasis affecting the transfer of QTLs for stress tolerance from the wild into the domesticated genepool, and trait expression in a wild relative may not be an adequate guide to its expression in a domesticated genetic background [37]. The trait will need to be further evaluated after 1–2 backcrosses into the domesticated background, probably of an elite breeding line chosen as the recurrent parent [37]. Survivalomics searches through the natural selection history of habitats that helped determine a landrace population, a searching backwards strategy to locate sources of stress tolerances. The important issue thus becomes the optimisation of genomic tools to enable breeding and selection for these traits.

GENOMICS AND CONTROL MECHANISMS FOR STRESS TOLERANCE

New genomic tools such the use of a very density of genetic markers such as single nucleotide polymorphisms (SNPs) can provide genome-wide selection (GWS) in conjunction with genotyping of each individual in a segregating population [46]. Each marker has a breeding value based on phenotypes and marker genotypes, and summing the breeding values of all markers provides a genomic estimated breeding value for selection of superior individuals. This approach can be used to select backcross progeny from wild × domestic crosses using bi-parental population mapping to detect rare alleles of moderate effect which may be subject to epistasis [47]. Genomewide selection with molecular markers were used to simulate selection directly in F2 population from a cross of adapted × exotic maize for introgression of favorable alleles, to illustrate applicability to plant breeding programs re population sizes, number of selection cycles, and replication across different crop environments and years [48]. A high density of molecular markers has been identified for the wild progenitor of chickpea, *Cicer reticulatum,* which will facilitate exploitation of this source of diversity for crop improvement [49]. The combining of these approaches with bioinformatics helps to integrate various disciplines; structural genomics, transcriptomics, physiology and plant breeding, towards a more targeted and efficient plant breeding plus elucidation of heterosis and epigenisis, but the costs may be prohibitive for public breeding programs for self-pollinated crops [50]. Yet as new technologies become increasingly cost effective it is expected that the new genetic tools will enhance but not replace conventional breeding.

Stress tolerant regulatory loci occur across various species as variants of candidate genes or even as direct analogues [38,51]. Molecular tools are available to assist identification of critical genes for drought and for heat tolerance from crop wild relatives [23], such as analysis of known candidate genes, and application of comparative functional genomics in a genome wide search. Hundreds of adaptation and regulatory genes are known to affect crop responses to heat and drought stresses. These may be expressed by differences in protein levels or protein sequences detectable by respective mRNA levels. Drought up-regulated genes conferring resistance include dehydrins (*Hordeum spontaneum*), and DREB in *Glycine soja* [38]. However a trait associated with drought or heat stress may actually be linked to a change in expression of another trait or even of a still unknown biochemical process. The molecular bases of various stress tolerance traits are not known in many species, despite documentation of a wide range of drought induced processes; e.g., anti-oxidant enzymes and compounds such as ascorbic acid, proline and mannitol which remove stress induced toxic reactive oxygen species molecules (ROS); osmoprotectant protein stabilisers in dehydrated plants; and heat shock HSPs/chaperones [38]. There is increased expression of heat protection genes in heat acclimated plants, than in non-acclimated plants exposed to a sudden heat spike. This corresponds to higher levels of mRNA of key heat protection genes. Adaptation to heat stress may depend more on the superior thermostability of key metabolic enzymes such as those for ROS removal, than the levels of stress protectant proteins.

Regulatory genes are important in the simultaneous changes of the mRNA levels of many genes such as for ROS removal [38]. Drought response regulators may include signalling proteins, transcription factors which bind to cis-elements in promoters, proteins for post transcriptional regulation of RNA, and enzymes for abscissic acid synthesis. ABA (abcisic acid) pathways are important in plant response to drought stress. A wild tomato (*Solanum chilense*) which is adapted to extremely dry environments had a divergent haplotype structure in a dehydrin gene downstream in the ABA pathway attributed to directional selection [52]. Heat response regulators, such as some HSPs, can affect various regulatory proteins to result in changes to leaf morphology and to flowering time. There are many forms of heat shock factors variously up-regulated by heat and by drought, and over-expression of some can improve heat tolerance. DREB genes may also be up-regulated by heat. Many complex gene expressions, regulatory systems and specific adaptation genes, interact with heat and drought stresses currently experienced by crops in many environments and which are expected to be more severe as climate changes. Trait associated genes may not possess the allelic variation governing the trait expression, but reflect the allelic variation in an upstream regulator [38]. However positional cloning of a gene/regulatory sequence associated with a target phenotype does not conclusively prove that this is the causal sequence variant for a trait [46].

Differentially expressed genes associated with genetic variation for heat or drought have been investigated through their EST sequences using large array analyses for genome wide searches, however this is likely to be super-ceded with next generation sequence (NGS) technologies. Both approaches enable comparative functional studies of genes for stress resistance. Candidate genes for drought/heat adaptive traits in the domesticated gene pools of crops, such as for osmotic adjustment, transpiration efficiency, photosynthetic maintenance and water soluble carbohydrates have been identified [38]. But for the crop wild relative gene pools, gene transcripts may differ widely from those for crops, and array technology based on crop transcriptome sequences may not be suitable for its wild relatives. Yet such difficulties may be overcome with the increasing availability of sequences with a high density of markers for some crop progenitors such as *Cicer reticulatum* for chickpea [49], and *Aegilops tauschii,* the source of the D genome in the hexaploid bread wheat [53]. Genetic differences between crops and wild relatives are too large for mapping of differentially expressed functional genes for a trait QTL. Yet precision is improved if the physiological and biochemical bases of the trait are known [54].

With NGS, whole genome or gene-space sequencing data are available for many crops, plus transcriptome sequence data to provide large numbers of ESTs [50]. NGS technologies are rapid and relatively cheap, and many parallel reactions can be multiplexed for which bioinformatic tools need further development. NGS resequencing is very useful for discovery of single nucleotide polymorphisms (SNPs) especially where a sequenced reference genome is available, and also for sequence assembly using bacterial artificial chromosomes for species with no reference genome available. NGS technologies can also be used for association mapping, wide crosses and alien introgression [55]. In particular NGS also allows additional molecular markers to be developed on a genome wide scale for wild germplasm enabling accuracy in tracing introgression of well defined segments from wild relatives. Knowledge of the gene region enables the use of NGS to screen germplasm collections to detect rare alleles in the domesticated gene pool and novel alleles in the wild genepools [46]. Thus previous limitations of genetic markers for traits in wild relatives can now be addressed with NGS facilitating exploitation of this wild diversity for crop improvement.

While NGS technologies hold much promise, and a further "Third generation technology" is being developed, there are complications for allopolyploid genomes such as bread wheat made up of three related sets of homoeologous chromosomes, and canola (*Brassica napus*) with genome complements from two diploid domestic Brassica crops. In wheat genomic analysis and construction of a reference genome is confounded by homoeologous syntenies and by repetitive transposon sequences in 75%–90% of the genome, with genic and low copy regions greatly interspersed by repeat sequences, thereby presenting challenges to genome sequencing [56]. Despite these difficulties the analysis of the bread wheat genome has now been achieved [57]. In Brassica there was an ancient triplication of the genome followed by diploidisation with very uneven loss of genome segments or genome fractionation. Thus the allopolyploid *B. napus* has complex presence-absence polymorphisms, and the subgenomes have a high level of plasticity with less structured patterns of gene loss compared with another reduced polyploid crop, maize [56]. Besides these complications, QTLs for stress and other trait expressions may not be stable across environments, and may be influenced by epistasis especially when there is a wide genetic distance between the QTL source and the introgressed genetic background [58]. Also such QTLs are affected by confounding expressions such as phenology affecting both drought response and grain yield relationships [46]. NGS and detailed data sets allow analysis of genic regions for tolerances of heat and other abiotic stress expressions at different stages of crop development, and better targeting of breeding opportunities.

SYNTHESIS OF PLANT BREEDING, GENOMICS AND GENOTYPE × ENVIRONMENT INTERACTION

Plant breeding addresses the challenge of combining together heritable traits for improved yield, resistances to endemic diseases and pests, and product quality traits, to suit the production system of the target environment. This target environment can be very wide when new varieties have general adaptation [59], with relative stability in performance across locations and years with differing crop seasons. One of the most difficult challenges is to select and evaluate promising selections across a suitable range of environments, with the almost inevitable occurrence of genotype × environment interaction in which ranking of entries alters in different locations. This challenge will be magnified by climate change with a predicted wider and more extreme fluctuation of seasons, and the requirement for new varieties by 2050 to posses resilience to extreme spikes in abiotic stresses [2,3], notably in the Indian sub-continent and sub-Saharan Africa, and to a lesser but significant extent in other temperate to sub-tropical latitudes. Survivalomics breeding to reduce risk in extremes of temperature and drought, plus selection for general adaptation to less reliable climates, will be key plant breeding challenges in the future. Wider genetic variation than currently available in domesticated crop gene pools will often be necessary to avoid replacement with alternative crops, with the largely unexploited genepools of crop wild relatives offering the most immediate prospects as exemplified in wheat and tomato [37,44]. Jordan *et al.* [60] describe a practical example of broadening the domesticated gene pool for quantitative traits, for a sorghum breeding program.

Early analyses of genotype × environment interaction by regression analyses suggested that patterns of wide and of specific adaptation found in a worldwide barley germplasm collection reflected the natural selection history of the respective germplasm [59]. Adaptation has been more precisely described with three way analysis of Chinese adzuki bean (*Vigna angularis*) landraces × multiple environments × multiple traits. Accessions were clustered into groups with distinctive patterns of adaptation across geographically and climatically diverse sites, which reflected differences in both trait phenotype combinations and in the geographic origin of accessions [61]. The adaptation characteristic of an accession was defined by the group to which it belonged. There were differences among the adzuki adaptation groups in levels of genotype × environment interaction across traits, which can also be interpreted as differences in plasticity [20,61]. Molecular analyses of Chinese pea landraces also displayed different genetic groups differing in geographic origins [62], and molecular analyses of faba bean landraces in China demonstrated clustering by geographic groups [63]. Variation in alleles for earliness in millet were associated with differences in rainfall, and earliness was correlated with both reduced height and shorter spikes, indicating a possible suite of traits for improved fitness and yield adaptation to drier environments [16]. It is possible that adaptation across environments can be defined genetically for many crops, explained in part by many underlying traits such as photoperiod response [64,65], and tolerances of soil moisture and of temperature stresses [24]. Thus if adaptation to target environments can be genetically characterized, possibly QTLs could be identified for various expressions of crop adaptation and plasticity [20]. The challenge is whether diverse genetic expressions for extreme drought and for heat tolerances can be selectively pyramided by strategic breeding to contribute the wide adaptation and plasticity for crops to cope with the predicted changes in world climate. It is important that development of such elite lines are planned now before 2050 in breeding programs tailored to each crop and target environment, to help crops adapt to climate change, rather than experience a lag in the development of appropriate research and development programs [11]. Overall there is the food security question of whether agriculture can meet the needs of future populations towards the end of the 21st century, particularly with the added complication of climate change.

CONCLUSIONS

New molecular tools such as NGS will assist in identifying the underlying components of stress tolerances in the domestic germplasm, and in the identification and introgression of novel stress genes from crop wild relatives. This is being systematically applied in the International Centre for Wheat and Maize Improvement (CIMMYT) wheat breeding program with managed field screening for tolerances of heat and drought stresses, to exploit both the domesticated and wild genepools [21,66]. For most crops the domestic germplasm has not been evaluated eco-geographically for identification of landrace collection sites likely to have had natural selection for tolerances of abiotic stresses, and likewise for the wild gene pools which have been little explored for novel genes for stress tolerances. With strategically directed research there may be enormous scope to select for adaptation to extreme abiotic stresses in the domesticated gene pools of crops, and in the largely untapped germplasm resources of related wild gene pools, for exploitation of survivalomics to assist crop adaptation to climate change.

REFERENCES

1. Intergovernmental Panel on Climate Change (IPCC). Summary for Policymakers. *Climate Change 2007: The Physical Science Basis. Contribution of Working Group I to the Fourth Assessment Report of the Intergovernmental Panel on Climate Change*; Solomon, S., Qin, D., Manning, M., Chen, Z., Marquis, M., Averyt, K.B., Tignor, M., Miller, H.L., Eds.; Cambridge University Press: Cambridge, UK, 2007.
2. Battisti, D.S.; Naylor, R.L. Historical warnings of future food insecurity with unprecented seasonal heat. *Science* 2009, *323*, 240–244.
3. Lobell, D.B.; Burke, M.B.; Tebaldi, C.; Mastrandrea, M.D.; Falcon, W.P.; Naylor, R.L. Prioritizing climate change adaptation needs for Food Security in 2030. *Science* 2008, *319*, 607–610.
4. Schafleitner, R.; Ramirez, J.; Jarvis, A.; Evers, D.; Gutierrez, R.; Scurrah, M. Adaptation of the Potato Crop to Changing Climates.

In *Crop Adaptation to Climate Change*; Yadav, S.S., Redden, R.J., Hatfield, J.L., Lotze-Campen, H., Hall, A.E., Eds.; Wiley-Blackwell: Chichester, UK, 2011; Chapter 11. pp. 287–297.

5. Singh, R.P.; Vara Prasad, P.V.; Sharma, A.K.; Raja Reddy, K. Impacts of High-Temperature Stress and Potential Opportunities for Breeding. In *Crop Adaptation to Climate Change*; Yadav, S.S., Redden, R.J., Hatfield, J.L., Lotze-Campen, H., Hall, A.E., Eds.; Wiley-Blackwell: Chichester, UK, 2011; Chapter 5.1. pp. 166–185.
6. Lafarge, T.; Peng, S.; Hasegawa, T.; William, P.; Quick, S.V.; Jagadish, K.; Wassmann, R. Genetic Adjustment to Changing Climates: Rice. In *Crop Adaptation to Climate Change*; Yadav, S.S., Redden, R.J., Hatfield, J.L., Lotze-Campen, H., Hall, A.E., Eds.; Wiley-Blackwell: Chichester, UK, 2011; Chapter 12. pp. 298–313.
7. Vadez, V.; Kholova, J.; Choudhary, S.; Zindy, P.; Terrier, M.; Krishnamurthy, L.; Ratna Kumar, P.; Turner, N.C. Responses to Increased Moisture Stress and Extremes: Whole Plant Response to Drought under Climate Change. In *Crop Adaptation to Climate Change*; Yadav, S.S., Redden, R.J., Hatfield, J.L., Lotze-Campen, H., Hall, A.E., Eds.; Wiley-Blackwell: Chichester, UK, 2011; Chapter 5.2. pp. 186–197.
8. Lotze-Campden, H. Climate Change, Population Growth, and Crop Production: An Overview. In *Crop Adaptation to Climate Change*; Yadav, S.S., Redden, R.J., Hatfield, J.L., Lotze-Campen, H., Hall, A.E., Eds.; Wiley-Blackwell: Chichester, UK, 2011; Chapter 1. pp. 1–11.
9. Redden, R.J.; Yadav, S.S.; Hatfield, J.L.; Prasanna, B.M.; Vasal, S.K.; Lafarge, T. The Potential of Climate Change Adjustment in Crops: A Synthesis. In *Crop Adaptation to Climate Change*; Yadav, S.S., Redden, R.J., Hatfield, J.L., Lotze-Campen, H., Hall, A.E., Eds.; Wiley-Blackwell: Chichester, UK, 2011; Chapter 24. pp. 482–494.
10. Turner, N.C.; Meyer, R. Synthesis of Regional Impacts and Global Agricultural Adjustments. In *Crop Adaptation to Climate Change*; Yadav, S.S., Redden, R.J., Hatfield, J.L., Lotze-Campen, H., Hall, A.E., Eds.; Wiley-Blackwell: Chichester, UK, 2011; Chapter 4. pp. 156–165.

11. Chapman, S.C.; Chakraborty, S.; Dreccer, M.F.; Howden, S.M. Plant adaptation to climate change-opportunities and priorities in breeding. *Crop Pasture Sci.* 2012, *63*, 251–268.

12. Diakité, L.; Sidibé, A.; Smale, M.; Grum, M. *Seed Value Chains for Sorghum and Millet in Mali. A State-Based System in Transition*; IFPRI Discussion Paper 00749; International Food Policy Research Institute: Washington, DC, USA, 2008.

13. Uprety, D.C.; Sirohi, G.S. Comparative study on the effect of water stress on the photosynthesis and water relations of triticale, rye and wheat. *J. Agron. Crop Sci.* 1987, *159*, 349–355.

14. Gowda, C.L.L.; Saxena, K.B.; Srivastava, R.K.; Upadhyaya, H.D.; Silim, S.N. Pigeonpea: From an Orphan to A Leader in Food Legumes. In *Biodiversity in Agriculture: Domestication, Evolution, and Sustainability*; Cambridge University Press: New York, NY, USA, 2011; pp. 362–373.ISBN 9780521764599.

15. Bennett, E. Adaptation in Wild and Cultivated Plant Populations. In *Genetic Resources in Plants—Their Exploration and Cultivation*; Frankel, O.H., Bennett, E., Eds.; IBP Handbook No 11; Blackwell Scientific Publications: Oxford, UK, 1970; pp. 115–129.

16. Vigouroux, Y.; Mariac, C.; De Mita, S.; Pham, J.-L.; Gerárd, B.; Sagnard, F.; Deu, M.; Chantereau, J.; Ali, A.; Ndjeung, J.; Thuillet, A.C.; Daidou, A.A.; Bezancon, G. Selection for early flowering crop associated climatic variations in the Sahel. *PloS One* 2011, 6, e19563.

17. Bao, S.; He, Y.; Zong, X.; Wang, L.; Li, L.; Enneking, D.; Rose, I.A.; Leonforte, T.; Redden, R.J.; Paull, J. Collection of pea (*Pisum. sativum*) and faba bean (*Vicia. faba*) germplasm in Yunnan. *Plant Genet. Resour. Newsl. FAO Bioversity* 2008, *156*, 11–22.

18. Chenbang, H.; Yujiao, L.; Kunlun, W.; Mingyi, Y.; Qinhua, F.; Yang, L.; Qingbiao, Y.; Jianping, G.; Rose, I.A.; Redden, R.J.; *et al.* Collecting and surveying landraces of pea (*Pisum. sativum*) and faba bean (*Vicia. faba*) in Qinghai province of China. *Plant Genet. Resour. Newsl. FAO Bioversity* 2008, *156*, 1–10.

19. Frankel, O.H.; Bennett, E. Genetic Resources-Introduction. In *Genetic Resources in Plants—Their Exploration and Conservation*; Frankel, O.H., Bennett, E., Eds.; IBP Handbook No 11; Blackwell Scientific Publications: Oxford, UK, 1970; pp. 7–17.

20. Mercer, K.L.; Perales, H. Evolutionary response of landraces to climatic change in centres of diversity. *Evol. Appl.* 2010, *3*, 480–493.

21. Trethowan, R.M.; Mahmood, T. Genetic Options for Improving Productivity of Wheat in Water-Linited and Temperature-Stressed Environments. In *Crop Adaptation to Climate Change*; Yadav, S.S., Redden, R.J., Hatfield, J.L., Lotze-Campen, H., Hall, A.E., Eds.; Wiley-Blackwell: Chichester, UK, 2011; Chapter 7. pp. 218–237.

22. Hijmans, R.J.; Cameron, S.E.; Parra, J.L.; Jones, P.G.; Jarvis, A. Very high resolution interpolated climate surfaces for global land areas. *Int. J. Climatol.* 2005, 25, pp. 1965–1978. Available online: http://www.worldclim.org/current (accessed on 7 May 2013).

23. Upadhyaya, H.D.; Dwivedi, S.L.; Ambrose, M.; Ellis, N.; Berger, J.; Smykal, P.; Bebouck, D.; Dumet, D.; Flavell, A.; Sharma, S.K.; *et al.* Legume genetic resources: Management, diversity assessment, and utilisation in crop improvement. *Euphytica* 2011, *180*, 27–47.

24. Li, L.; Redden, R.J.; Zong, X.; Berger, J.D.; Bennett, S.J. Ecogeographic analysis of pea collection sites from China to determine potential sites with abiotic stresses. *Genet. Resour. Crop Evol.* Available online: http://www.springerlink.com/openurl.asp?genre=article&id=doi:10.1007/s10722-013–9955-6 (accessed on 7 May 2013).

25. Islam, F.; Beebe, S.; Munoz, M.; Tohme, J.; Redden, R.J.; Basford, K.E. Using molecular markers to assess the effect of introgression on quantitative attributes of common bean in the Andean gene pool. *Theor. Appl. Genet.* 2004, *108*, 243–252.

26. Berger, J.D.; Mackay, M.C.; Street, K.A.; Konopka, J.; Adhikari, K.; Clarke, H.J.; Sandhu, J.S.; Nayyar, H. Emerging Opportunities for Agriculture: Investigating Plant Adaptation by Characterizing Germplasm Collection Habitats. In Proceedings of the 14th Australian Agronomy Conference, Global Issues, Paddock Action, 5395, Adelaide, Australia, 21–25 September 2008.

27. Ehlers, J.D.; Hall, A.E. Heat tolerance of contrasting cowpea lines in short and long days. *Field Crops Res.* 1998, *55*, 11–21.

28. Hall, A.E. Breeding Cowpea for Future Climates. In *Crop Adaptation to Climate Change*; Yadav, S.S., Redden, R.J., Hatfield,

J.L., Lotze-Campen, H., Hall, A.E., Eds.; Wiley-Blackwell: Chichester, UK, 2011; Chapter 15. pp. 340–355.

29. Abbo, S.; Berger, J.; Turner, N.C. Evolution of cultivated chickpea: Four bottlenecks limit diversity and constrain adaptation. *Funct. Plant Biol.* 2003, *30*, 1081–1087.
30. Mayr, E. *Systematics and the Origin of Species*; Columbia Univ. Press: New York, NY, USA, 1942.
31. Ladizinsky, G. The Course of Reducing and Maintaining Genetic Diversity under Domestication. In *Plant Evolution under Domestication*; Kluwer Academic Publishers: Dortrecht, The Netherlands, 1998; Chapter 3. pp. 113–126.
32. Ladizinsky, G. Origins of Agriculture. In *Plant Evolution under Domestication*; Kluwer Academic Publishers: Dortrecht, The Netherlands, 1998; Chapter 1. pp. 1–60.
33. Meyer, R.S.; DuVal, A.E.; Jensen, H.R. Patterns and processes in crop domestication: An analysis of 203 global food crops. *New Phytol.* 2013, *196*, 29–48.
34. Vardi, A.; Zohary, D. Introgression in wheat via triploid hybrids. *Heredity* 1967, *22*, 541–560.
35. Hancock, J.F. The Dynamics of Plant Domestication. In *Plant Evolution and the Origin of Crop Species*, 3rd ed.; CABI: Wallingford, UK, 2012; Chapter 7. pp. 114–131.
36. Nguyen, T.T.; Taylor, P.W.J.; Redden, R.J.; Ford, R. Mining resistance to *Ascochyta. rabiei* in a wild *Cicer.* germplasm collection. *Aust. J. Exp. Agric.* 2005, *45*, 1291–1296.
37. Tanksley, S.D.; McCouch, S.R. Seed banks and molecular maps: Unlocking genetic potential from the wild. *Science* 1997, *227*, 1063–1066.
38. Xue, G.-P.; McIntyre, C.L. Wild Relative and Transgenic Innovation for Enhancing Crop Adaptation to Warmer and Drier Climate. In *Crop Adaptation to Climate Change*; Yadav, S.S., Redden, R.J., Hatfield, J.L., Lotze-Campen, H., Hall, A.E., Eds.; Wiley-Blackwell: Chichester, UK, 2011; Chapter 27. pp. 522–545.
39. Gur, A.; Zamit, D. Unused natural variation can lift yield barriers in plant breeding. *PLoS Biol.* 2004, *2*, 1610–1615.
40. Ladizinsky, G. Genetic Resources for Future Crop Evolution. In *Plant Evolution under Domestication*; Kluwer Academic

Publishers: Dortrecht, The Netherlands, 1998; Chapter 7. pp. 209–222.

41. Redden, R.J.; Berger, J.D. History and Origin of Chickpea. In *Chickpea Breeding & Management*; Yadav, S.S., Redden, R., Chen, W., Sharma, B., Eds.; CABI: Wallingford, UK, 2007; Chapter 1. pp. 1–13.
42. Ladizinsky, G. Weeds and Their Evolution. In *Plant Evolution under Domestication*; Kluwer Academic Publishers: Dortrecht, The Netherlands, 1998; Chapter 5. pp. 156–159.
43. Ellstrand, N.C.; Prentice, H.C.; Hancock, J.F. Gene flow and introgression from domesticated plants into their wild relatives. *Annu. Rev. Ecol. Syst.* 1999, *30*, 539–563.
44. Haijar, R.; Hodgins, T. The use of wild relatives in crop improvement: A survey of developments over the last 30 years. *Euphytica* 2007, *156*, 1–13.
45. Erskine, W.; Sarker, A.; Ashraf, M. Reconstructing an ancient bottleneck of the movement of the lentil (*Lens culinaris* ssp. *culinaris*) into South Asia. *Genet. Resour. Crop Evol.* 2010, *58*, 373–381.
46. Langridge, P.; Fleury, D. Making the most of the 'omics' for crop breeding. *Trends Biotechnol.* 2011, *29*, 33–40.
47. Rafalski, J.A. Association genetics in crop improvement. *Curr. Opin. Plant Biol.* 2010, *13*, 174–180.
48. Bernado, R. Genomewide selection for rapid introgression of exotic germplasm in maize. *Crop Sci.* 2009, *49*, 419–425.
49. Jhanwar, S.; Priya, P.; Garg, R.; Parida, S.K.; Tyagi, A.K.; Jain, M. Transcriptome sequencing of wild chickpea as a rich redource for marker development. *Plant Biotechnol. J.* 2012, *10*, 690–702.
50. Varshney, R.K.; Graner, A.; Sorrells, M.E. Genomics-assisted breeding for crop improvement. *Trends Plant Sci.* 2005, *10*, 621–630.
51. Weller, J.L.; Chee liew, L.; Hecht, V.F.G.; Rajandan, V.; Laurie, R.E.; Ridge, S.; Wenden, B.; Vander Schoor, J.K.; Jaminon, O.; Blassiau, C.; *et al.* A conserved molecular basis for photoperiod adaptation in two temperate legumes. *Proc. Natl. Acad. Sci. USA* 2012, *109*, 21158–21163.

52. Xia, H.; Camus-Kulandaivelu, L.; Dtephan, W.; Tellier, A.; Zhang, Z. Nucleotide diversity patterns of local adaptation at drought related candidate genes in wild tomatoes. *Mol. Ecol.* 2010, *19*, 4144–4154.

53. You, F.M.; Huo, N.; Deai, K.R.; Gu, Y.Q.; Luo, M.; McGuire, P.; Dvorak, J.; Anderson, O. Annotation-based genome-wide SNP discovery in the large and complex *Aegilops. tauschii* genome using next-generation sequencing without a reference genome. *BMC Genomics* 2011, *12*, 59–84.

54. Xue, G.P.; McIntyre, C.L.; Jenkins, C.L.D.; Glassop, D.; van Herwaarden, A.F.; Shorter, R. Molecular dissection of variation in carbohydrate metabolism related to water soluble carbohydrate accumulation in stems of wheat (*Triticum. aestivum* L.). *Plant Physiol.* 2008, *146*, 441–454.

55. Varshney, R.K.; Nayak, S.N.; Gregory, D.; May, G.D.; Jackson, S.A. Next-generation sequencing technologies and their implications for crop genetics and breeding. *Trends Biotechnol.* 2009, *27*, 522–530.

56. Edwards, D.; Batley, J.; Snowden, R.J. Assessing complex crop genomes with next-generation sequencing. *Theor. Appl. Genet.* 2013, *126*, 1–11.

57. Brenchley, R.; Spannagl, M.; Pfeifer, M.; Barker, G.L.A.; D'Amore, R.; Allen, A.; McKenzie, N.; Kramer, M.; Kerhornou, A.; Bolser, D.; *et al.* Analysis of the bread wheat genome using whole-genome shot-gun sequencing. *Nature* 2012, *421*.

58. Redden, R. The effect of epistasis on chromosome mapping of quantitative characters in wheat. II. Agronomic characters. *Aust. J. Agric. Res.* 1991, *42*, 335–345.

59. Finlay, K.W.; Wilkinson, G.N. The analysis of adaptation in a plant breeding program. *Aust. J. Agric. Res.* 1963, *14*, 742–754.

60. Jordan, D.R.; Mace, E.S.; Cruickshank, A.W.; Hunt, C.H.; Henzell, R.H. Exploring and exploiting genetic variation from unadapted sorghum germplasm in a breeding program. *Crop Sci.* 2011, *51*, 1444–1457.

61. Redden, R.J.; Basford, K.E.; Kroonenberg, P.M.; Amirul Islam, F.M.; Ellis, R.; Wang, S.; Cao, Y.; Zong, X.; Wang, X. Variation in

adzuki bean (*Vigna. angularis*) germplasm grown in China. *Crop Sci.* 2009, *49*, 771–782.

62. Zong, X.; Redden, R.J.; Liu, U.Q.; Wang, S.; Guan, J.; Liu, J.; Xu, Y.; Liu, X.; Gu, J.; Yan, L.; *et al.* Analysis of a diverse global *Pisum.* sp. Collection and comparison to a Chinese local *P. sativum* collection with microsatellite markers. *Theor. Appl. Genet.* 2009, *118*, 193–204.
63. Wang, H.; Zong, X.; Guan, J.; Yang, T.; Sun, X.; Ma, Y.; Redden, R. Genetic diversity and relationship of global faba bean (*Vicia. faba* L.) germplasm revealed by ISSR markers. *Theor. Appl. Genet.* 2012, *124*, 789–797.
64. Wang, S.M.; Redden, R.J.; Hu, J.P.; Desborough, P.J.; Lawrence, P.L.; Usher, T. Chinese adzuki bean germplasm: 1. Evaluation of agronomic traits. *Aust. J. Agric. Res.* 2001, *52*, 671–681.
65. Desborough, P.; Lawrence, P.; Redden, R.; Xuxiao, Z. Characterisation of Response to Temperature and Photoperiod in a Core Collection of Adzuki Bean from China. In Plant Breeding for the 11th Millennium; Proceedings of the 12th Australasian Plant Breed Conference, Perth, Australia, 15–20 September 2002; McComb, J., Ed.; pp. 565–568.
66. Lopes, M.S.; Reynolds, M.P.; Jalal-Kamali, M.R.; Moussa, M.; Feltaous, Y.; Tahir, I.S.A.; Barma, N.; Vargas, M.; Mannes, Y.; Baum, M. The yield correlations of selectable phenotypic traits in a population of advanced spring lines grown in warm and drought environments. *Field Crops Res.* 2012, *128*, 129–136.

Chapter 6

Effects of Previous Crop Management, Fertilization Regime, and Water Supply on Potato Tuber Proteome and Yield

Catherine Tétard-Jones[1], Martin G. Edwards[2], Leonidas Rempelos[1], Angharad M.R. Gatehouse[2], Mick Eyre[1], Stephen J. Wilcockson[1], and Carlo Leifert[1]

[1]School of Agriculture, Food and Rural Development, Molecular Agriculture Group, Nafferton Ecological Farming Group, Newcastle University, Nafferton Farm, Stocksfield, UK

[2]School of Biology, Molecular Agriculture Group, Newcastle University, Newcastle-Upon-Tyne, NE1 7RU, UK

ABSTRACT

There is increasing concern about the sustainability and environmental impacts of mineral fertilizer use in agriculture. Increased recycling of nutrients *via* the use of animal and green manures and fertilizers made from domestic organic waste may reduce reliance on mineral fertilizers. However, the relative availability of nutrients (especially nitrogen) is lower in organic compared to mineral fertilizers, which can result in significantly lower yields in nutrient demanding crops such as potato. It is therefore important to gain a better understanding of the factors affecting nutrient use efficiency (yield per unit fertilizer input) from organic fertilizers. Here we show that (a) previous crop management (organic *vs*. conventional fertilization and crop protection regimes), (b) organic fertilizer type and rate (composted cattle manure *vs*. composted chicken manure pellets) and (c) watering regimes (optimized and restricted) significantly affected leaf chlorophyll content, potato tuber N-concentration, proteome and yield. Protein inference by gel matching indicated several functional groups significantly affected by previous crop management and organic fertilizer type and rate, including stress/defense response, glycolysis and protein destination and storage. These results indicate genomic pathways controlling crop responses (nutrient use efficiency and yield) according to contrasting types and rates of organic fertilizers that can be linked to the respective encoding genes.

INTRODUCTION

There is increasing concern about the sustainability of intensive crop production systems and in particular their dependence on mineral (NPK) fertilizer inputs [1]. The manufacture of mineral N-fertilizer requires large energy inputs and is estimated to account for up to 10% of greenhouse gas (GHG) emissions from agriculture globally [2]. The high mineral fertilizer inputs used in crops such as potato causes significant N and P losses (through leaching and run-off) from agricultural land, which contributes to pollution and eventual eutrophication of fresh water and marine ecosystems [1,3]. There is also growing concern about the future availability of P, K and all other minerals used as fertilizer that are mined from finitely available natural deposits. For example, currently known deposits of phosphorus are estimated to last for only 30–100 years, depending on the rate of usage [4,5,6]. This combined with the increasing cost of mineral fertilizers has resulted in a re-evaluation of using organic fertilizers (animal and green manures, and organic waste composts) to replace or minimize the use of mineral fertilizer [7,8,9].

The use of organic fertilizers can increase the organic matter/carbon content, structural stability, biological activity and invertebrate biodiversity of soils, and reduce nitrogen and phosphorus leaching/run off [7,8,9,10,11]. However, at the same total N-input level, yields were often found to be lower with organic fertilizers compared with standard mineral fertilization regimes. Current environmental regulations limit applications of organic fertilizers to the equivalent of 170 kg N ha^{-1} $year^{-1}$ in order to minimize nitrate leaching losses [12]. For potato, the use of composted cattle manure at this input level was shown to result in (a) reduced concentrations of certain mineral nutrients (especially N and K) in plant tissues and leaf chlorophyll levels and (b) changes in protein expression profiles in tubers and most importantly, there was an increased expression of proteins involved in stress regulation [13,14]. This is thought to be due to (a) the virtual absence of plant available forms of nitrogen (NH_4^+ and NO_3^-) in composted manure and (b) insufficient release of plant available nutrients via mineralization of organic matter in soils; mineralization of soils is positively correlated to soil biological activity which is affected by previous soil management (e.g. increased by regular organic fertilizer inputs) and soil water availability [15,16]. This view is supported by the finding that potatoes fertilized with chicken manure pellets (which have a higher content of plant available N than composted manure) achieved higher yields than those grown in composted cattle manure treated soils [17].

It was therefore hypothesized that the N-use efficiency (potato yield per unit N-input) from organic fertilizer may be increased by (a) growing potato crops in soils which had regular organic matter inputs in previous years, (b) switching to organic fertilizers with a higher content of plant available N (e.g., chicken manure pellets) and (c) by optimizing soil water content [18,19], However, there is a lack of published studies to test these hypotheses. The development of both innovative soil and crop management practices and breeding strategies aimed at improving nutrient (especially N, P and K) use efficiency and crop yields from organic fertilizer requires a greater understanding of plant response at the molecular level. Protein profiling (proteomics) is a reliable technique for the detection of gene products expressed within plant tissue, enabling a comparison of differential expression in gene products across cropping systems [20,21]. This is a functional genomics approach that can identify proteins that are linked to causal genes, and enable the potential development of functional molecular markers for crop improvement in nutrient use efficiency [22].

In this study we assessed the effects of (a) contrasting organic fertilizers (composted cattle manure and composted chicken manure pellets), and fertilizer input levels (85 and 170 kg N ha^{-1}); (b) previous crop management (soils managed to organic or conventional farming standards for 4 years) and (c) contrasting water supply, on leaf chlorophyll content, tuber yield, N concentration, and protein profiles.

MATERIALS AND METHODS

Experimental Design

In 2005, a factorial pot experiment comparing contrasting fertilization and watering regimes was conducted at Close House Experimental Station (University of Newcastle). Seed potatoes of the variety Santé were planted (in April 2005) in 40 liter pots (one tuber per pot) containing soil collected from a field previously managed to either (i) conventional or (ii) organic farming practices. The two soils were collected from adjacent fields both with uniform sandy loam, at Nafferton Experimental Farm (University of Newcastle). The conventionally managed field had previously been sown with winter oilseed-rape, treated with aldicarb (Temik) for the control of soil pests, and received mineral fertilizer inputs. The soil from the field managed to organic farming standards had been cropped with a grass/red clover ley for four years and received applications of composted cattle manure. For each soil type, a factorial experiment was designed to compare fertilization type (composted cattle manure and composted chicken manure pellets), fertilization level (85 kg N ha^{-1} and 170 kg N ha^{-1}, plus a control—0 N) and watering regime (restricted and optimized), producing 10 treatments per soil type (Table 1). The two fertilization types contained different amounts of P and K kg^{-1} N (Table 2). Plants were observed daily throughout the whole period of growth, and water was applied at the rate of 2 liters of water per application to individual pots for the optimum treatment whenever plants began to show the first signs of stress/wilting. For the restricted watering treatment, 1 liter of water per pot was applied when the optimum treatment plants were watered and observations confirmed that at the time that water was applied, these plants were more stressed than those receiving optimum watering. A randomized block design was used, with the treatments randomized within each of 4 replicate blocks and the experiment was repeated.

Table 1: Experimental design to illustrate the combination of watering regimes, previous crop management (soil from conventional or organic managed fields), and fertilization types and levels (low input = 85 kg N ha^{-1}; High input = 170 kg N ha^{-1}). Asterisks indicate the eight treatments from which tubers samples were analyzed by 2D gel electrophoresis

	Restricted watering	Optimum watering
Conventional previous crop management		
Control	No fertilization	No fertilization *
Composted cattle manure	Low input	Low input
	High input	High input *
Composted chicken manure pellets	Low input	Low input
	High input	High input *
Organic previous crop management		
Control	No fertilization	No fertilization *
Composted cattle manure	Low input	Low input *
	High input	High input *
Composted chicken manure pellets	Low input	Low input *
	High input	High input *

Table 2: Inputs of major nutrients applied (kg ha^{-1}) in composted cattle manure and chicken manure pellets at two application levels: low input = 85 kg N ha^{-1}; High input = 170 kg N ha^{-1}

	Composted cattle manure		Composted chicken manure pellets	
	Low input	High input	Low input	High input
C	931	1863	745.7	1491.4
Total N	85	170	85	170
NH_4^+	1.0	2.0	20.2	40.4

NO_3^-	1.4	2.8	0.35	0.70
Organic N	82.6	165.2	64.5	129.0
P	27.2	54.4	24.3	48.6
P_2O_5	62.6	125.2	57.4	114.8
K	79.7	159.4	48.7	97.3
K_2O	95.6	191.2	57.4	114.8

Chlorophyll Content

Leaf chlorophyll content measurements were estimated at two different times (24 July and 3 August 2005) using a SPAD meter (SPAD-502, Konica Minolta Sensing Inc.-Japan) to determine the nitrogen status of plant leaves on 24 June and 3 August. These dates corresponded to 41 days (BBCH scale foliar growth stage 55—inflorescence emergence, tuber growth stage 40—tuber formation) and 81 days (BBCH scale foliar growth stage 80—berry formation, tuber growth stage 46—main period of tuber growth) after emergence [23].

Yield Assessments

Tuber yield was assessed at the end of October. The number and weight of tubers in each of three size categories: <45 mm, 45–65 mm, and 65–85 mm and totals were recorded. Tuber dry matter content was determined on a composite, fresh sub-sample of tubers of approximately 150 g taken from the different size grades, which was dried in a forced-draught oven at 80 °C for 48 h. Tuber nitrogen composition was analyzed for total N by Dumas combustion according to the application notes provided by the instrument's producer (LECO TruSpec Automated C/N Analyzer, LECO Corporation, USA).

Protein Profiling

A subset of the treatments included in the pot experiments was selected for protein profiling of the potato tubers (Table 1). Only tubers from plants under the optimum irrigation regime were used. Three fertilization treatments were selected (0 N, cattle manure compost at 170 kg N/ha and composted chicken manure pellets at 170 kg N/ha) for plants grown in soils that were previously under conventionally management, while all five fertilization treatments were selected (composted cattle manure at 85 kg N $ha^{-1}$and 170 kg N ha^{-1}, composted chicken manure pellets at 85 kg N ha^{-1} and 170 kg N ha^{-1}, and the control—0 N ha^{-1}) for plants grown in soils that were previously under organic management. An average size tuber was randomly collected from each of the selected treatments, washed in double distilled water then sliced, lyophilized, and milled into a fine powder for protein extraction.

Protein Extraction and 2-DE

Proteins were extracted from approximately 100 mg freeze dried material (typically equivalent to 0.9–1.0 g fresh tissue) using a phenol extraction method as previously described [24]. Total soluble protein content was measured using a 2D-Quant kit (Amersham). For each sample, 200 µg was precipitated from the phenol extraction buffer using a 2D-Clean Up kit (Amersham). The protein pellet was then resolubilized in 85 µL De-Streak buffer with 0.5% pH 4–7 IPG buffer. Protein extracts from each sample within each treatment were then pooled to give a total protein load of 800 µg per 2-DE gel (biological replicates: 8, technical replicates: 6).

First-dimension isoelectric focusing was performed using 18-cm isoelectric focusing (IEF) strips (Amersham Biosciences) with a linear pH range of 4–7 in an Ettan IPGPhor isoelectric focusing system. The IEF strips were rehydrated overnight with total protein, then focusing was run using the following conditions: from 0 to 500 V in 1 h, from 500 to 1000 V in 1 h, from 1000 V to 8000 V in 3 h, finally 8000 V until 36,000 Vh. After focusing, the strips were stored at −70 °C then equilibrated at room temperature in equilibration buffer (6 M urea, 50 mM Tris-HCl, pH 8.8, 30% (v/v) glycerol, 2% (w/v) SDS) with 1% (w/v) DTT for 10 min. This buffer was decanted and replaced with equilibration buffer containing 2.5% (w/v) iodoacetamide for a further 10 min. The second dimension was run in the Ettan DALTsix system with 19 × 23 cm homogeneous 12% SDS-PAGE gels, according to standard protocols [25]. The gels were run with constant 25- to 30-mA current overnight. The gels were fixed and stained with colloidal coomassie G-250 (Candiano) overnight and destained in double distilled water. Gels were scanned with a flatbed scanner to give 16 bit resolution images for subsequent analysis.

Image Analysis

The 2-DE gel images were analyzed using Progenesis SameSpots (Nonlinear Dynamics). Protein spot volumes were normalized using the total spot volume normalization method, whereby each spot on a gel image is expressed relative to the total volume of all spots on that image. This method of normalization is designed to minimize differences in spot volume caused by gel–gel variation. Overall, 302 distinct spots were matched across the 2-DE gels, and their normalized spot volumes exported from Progenesis for further statistical analysis.

Statistical Analysis

The effects of previous crop management regime, fertilization type and level, and watering regime on SPAD measurements, the potato tuber yield, and protein expression were assessed using ANOVA derived from linear mixed-effects models [26]. In an initial four factor ANOVA, SPAD data, tuber yield, tuber N concentration and percentage dry matter were analyzed with experiment, watering regime, soil type and fertilization regime as fixed factors, and experiment, replicate blocks, and watering regime as random factors [27]. We found that experiment was not a significant factor affecting any of the traits (data not shown). Therefore, the data was re-analyzed using 3 factor ANOVA (without experiment as a fixed factor). For the normalized protein spot data, the effects of soil and fertilization regime (at the high input level and optimum watering regime) were tested in a two way ANOVA with soil and fertilization regime as fixed factors, and replicate blocks as random factors. All analyses were carried out using the nlme library in the R statistical environment [28]. Residual normality was assessed using the qqnorm function in R [27], with no data showing serious violations from normality.

The relationships of tuber yield components and tuber protein spots to the agronomic regimes were investigated with redundancy analyses (RDA), using the CANOCO package [29]. Automatic forward selection of the agronomic regimes within RDA was used and their significance calculated using Monte Carlo permutation tests.

Protein Inference by Gel Matching

Protein spots whose abundance was significantly affected by Previous Crop Management (PCM), and/or fertilization type and/or rate were selected for inference by gel matching, in order to assign an inferred function. This method has been used previously to infer function based on gel matching in previous studies when protein identification was not possible [30]. Gel matching in the present study was based on a reference gel image, created by Progenesis Same Spots (Nonlinear Dynamics) during the image analysis procedure. The reference gel image was used to compare the migration of spots with three previously published potato tuber proteome maps that had been generated by the same 2-DE gel methods to separate protein spots [14,31,32]. Of these three studies, one of them used the same potato variety as used in the present study—Santé [14]. Proteins were inferred by matching their spatial pattern between the reference gel image generated in this study to the published maps. A grid was created onto the reference gel image and published maps to aid gel matching. Protein spots had to meet several criteria to meet confidence of inference: (i) proteins spots were characteristic, *i.e.*, present on the reference gel image and had low variance within treatments, (ii) protein spots were one of several in an area of the gel that formed a characteristic pattern replicable over gels in the present study and the previously published gel map used for matching. Protein spots that did not meet these criteria were discarded. For many protein spots, there was a positive match between all three studies, giving us confidence in this gel matching approach. We have cited multiple references for protein spots that were matched and identified in multiple studies. However, a large proportion of the matched spots only have one or two of the three citations; in these cases the protein spots were only identified in one or two of the published studies even if they were matched across gel maps in all three studies. In the cases where we matched the same protein spot across two or three of the published maps and that had been identified in each study, the protein identifications were the same (*i.e.*, conflicting protein identities were not found), thus further increasing our confidence in this gel matching approach.

RESULTS

Effects of Previous Crop Management, Fertilization Regime, and Watering Supply on Potato Agronomic Traits

Chlorophyll Content of Potato Leaves

SPAD measurements were taken as an estimate of chlorophyll content of potato leaves at 41 and 81 days after emergence; chlorophyll content is known to be closely linked to N status of potato plants. Chlorophyll content was lower at the later assessment date, 81 days after emergence compared to 41 days after emergence (supplementary Table S1).

Two factors in the analysis (watering regime and fertilization regime) significantly affected chlorophyll content at both dates, and previous crop management had a significant effect at 41 days after emergence ($p < 0.05$, supplementary Table S1). Chlorophyll content was significantly higher when the restricted watering regime was used ($p < 0.001$), and when plants were grown in PCM_{con} (conventional Previous Crop Management, at 41 days after emergence only, $p < 0.05$). Chlorophyll content was highest in soils fertilized with composted chicken manure pellets at the higher input rate (170 kg N ha^{-1}) followed by composted chicken manure pellets at 85 kg N ha^{-1} then the two cattle manure compost input treatments and lowest in the non-fertilized control.

A significant interaction between watering regime and previous crop management was detected at both dates ($p < 0.05$, supplementary Table S2). With restricted watering, potato leaves had a lower amount of chlorophyll when grown in PCM_{org} (organic Previous Crop Management) compared to PCM_{con}. In contrast, when the optimized watering regime was used, higher chlorophyll content was detected in plants grown in PCM_{org} when compared to plants grown in PCM_{con}.

Potato Tuber Yield

Total tuber fresh yield was significantly ($p < 0.001$) affected by each of the three experimental factors: previous crop management (soil previously managed to organic or conventional fertilization, crop protection and tillage standards), fertilization regime (type and level of fertilizer input applied before planting potato), and watering regime (restricted and optimized) (Table 3). Yield recorded for two of the three tuber size categories (65–85 mm and <45 mm) was also significantly affected by previous crop management, fertilization regime, and watering regime ($p < 0.05$, supplementary Table S3). In contrast, only previous soil management significantly affected dry matter content of tubers ($p < 0.0001$). Growing potatoes in soils that were previously under conventional management (PCM_{con}) led to significantly higher total tuber yield, but lower dry matter content in tubers when compared to potato grown in soils that were previously managed to organic farming standards (PCM_{org}) (Table 3). There were no effects of interactions between factors on tuber yield components.

Table 3: Effects of previous crop management (PCM), fertilization regime (FR), and watering regime (WR) on total tuber yield, and tuber N concentration

Factor	Tuber number plant^{-1}	Tuber fresh weight (g) plant^{-1}	N concentration (%)	Dry matter (%)
Means (± SE)				
WR				
Restricted	13.2 ± 0.4	575.3 ± 10.9	1.16 ± 0.02	23.8 ± 0.2
Optimized	14.1 ± 0.3	733.4 ± 12.0	1.03 ± 0.02	23.1 ± 0.3
PCM				
Conventional soil	15.2 ± 0.4	691.5 ± 14.8	1.09 ± 0.03	22.7 ± 0.2
Organic soil	12.1 ± 0.3	617.1 ± 12.8	1.10 ± 0.02	24.1 ± 0.2
FR				
Control	12.3 ± 0.4 c	563.9 ± 17.6 d	1.02 ± 0.03	23.5 ± 0.4 ab
Composted cattle manure, 85 kg N ha^{-1}	12.9 ± 0.6 bc	620.0 ± 16.4 c	1.09 ± 0.04	23.4 ± 0.4 ab
Composted chicken manure pellets, 85 kg N ha^{-1}	13.2 ± 0.7 bc	667.7 ± 21.8 b	1.09 ± 0.04	23.6 ± 0.4 ab
Composted cattle manure, 170 kg N ha^{-1}	14.1 ± 0.5 b	654.6 ± 19.7 bc	1.14 ± 0.03	22.8 ± 0.5 b

Composted chicken manure pellets, 170 kg N ha^{-1}	15.7 ± 0.6 a	765.4 ± 22.9 a	1.14 ± 0.04	23.9 ± 0.3 a
ANOVA				
Main effects				
WR	ns	<0.0001	0.0024	0.0615
PCM	<0.0001	<0.0001	ns	<0.0001
FM	<0.0001	<0.0001	0.0607	ns
Interaction effects				
WR × PCM	ns	ns	0.0062	0.0777
WR × FR	ns	0.0711	ns	ns
PCM × FR	ns	ns	0.0452	0.0687
WR × PCM × FR	ns	ns	ns	ns

Nitrogen Content of Potato Tubers

The nitrogen content of tubers was significantly affected by watering regime (Table 3). Tuber N-content was significantly higher when the restricted watering regime was used ($p < 0.01$). There was also a significant 2-way interaction between watering and previous crop management (supplementary Table S2). When watering was restricted, tuber N concentration was higher in tubers grown in PCM_{con}. However, when watering was optimized, tuber N concentration was higher in tubers grown in PCM_{org}(supplementary Table S2). There was also a significant 2-way interaction between previous crop management and fertilization treatments (supplementary Table S4). In PCM_{con} the tuber N-content significantly increased when composted chicken manure input was doubled, while there were no significant differences between the two input levels in $PCM_{org.}$

Associations between Agronomic Factors and Tuber Yield and Yield Components

The biplot in Figure 1 shows the relationship between agronomic factors (mineral nutrient input from organic fertilizers, watering regime and previous soil management) and yield/yield components. Axis 1 explained most (39%) of the variation and axis 2 a further 3%. Most additional variance was explained by watering regime (F = 36.7, P = 0.002), available N input (F = 26.1, P = 0.002) and previous soil management (F = 11.8, P = 0.002), but C-input (F = 12.5, P = 0.022) and experiment (F = 12.5, P = 0.004) were also significant drivers.

Nutrient input levels (NH_4/NO_3-N, organic-N, P, C and K) and tuber yield components showed strong positive association along the positive axis 1 for the numbers and fresh weight of total and large tubers and weaker positive associations for the numbers and fresh weight of smaller tubers. In contrast, there were weak negative associations between nutrient input levels and dry matter and N concentrations in tubers along the negative axis 1 (Figure 2). Tuber yield components showed very similar associations with optimized watering regimes (WR2), previous conventional soil management (PCM_{con}) and experiment 2. Experiment 2 was also positively associated with leaf chlorophyll content (SPAD) along the negative axis 2.

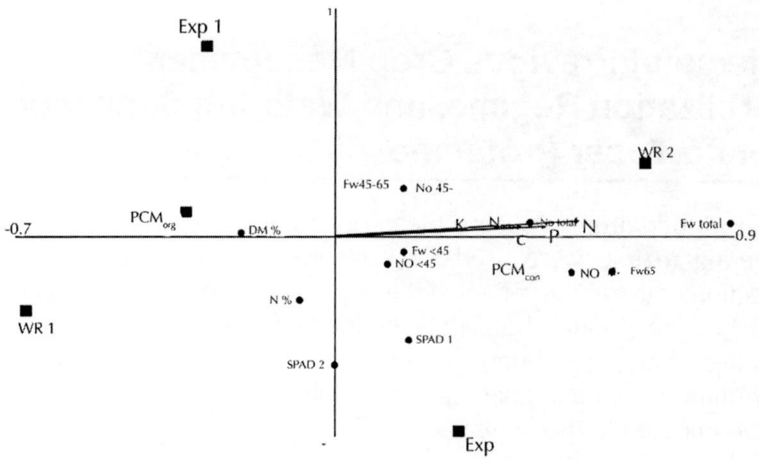

Figure 1: Bi-plot derived from redundancy analysis showing the relationship between potato tuber yield components and agronomic factors (categorical variables: organic and conventional previous soil management (PCM_{org}, PCM_{con}), restricted and optimized watering regime (WR1, WR2), experiments 1 and 2 (Exp 1, Exp 2) indicated by black squares; continuous variables: C, P, K, organic-N [N_{org}], available N [N_{av}]). DM % = percentage dry matter of tubers, FW = tuber fresh weight and No. = number of tubers (in each tuber size category).

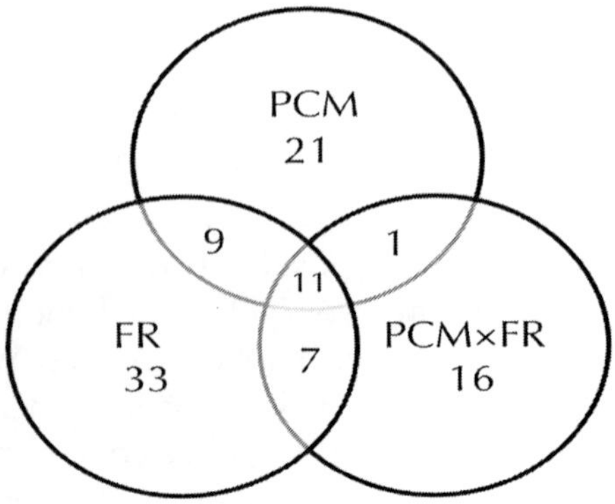

Figure 2: Venn diagram showing the number of protein spots significantly affected by previous crop management (PCM), fertilization regime (FR—composted chicken manure pellets *vs.* composted cattle manure at 170 kg N ha^{-1}), and the PCM × FR interaction effect ($p < 0.01$).

Effects of Previous Crop Management, Fertilization Regime, and Watering Supply on Potato Tuber Proteome

Potato tubers from a subsample of treatments within the optimized watering regime were used for proteomics analysis (Table 1). The selection of treatments enabled the effect of previous soil management and fertilizer inputs (but not watering regimes) on protein profiles (relative change in abundance of different proteins) in tubers to be determined. For the higher input level the effect of fertilizer input type was compared to the untreated control in both PCM_{org} and PCM_{con}. In addition we determined the effect of fertilizer input type and level in PCM_{org} only. A total of 302 protein spots were detected in the tuber proteome using 2D electrophoresis. Normalized spot volumes from each treatment and technical replicates were analyzed by ANOVA.

When the effect of fertilizer input type (composted chicken manure pellets and composted cattle manure applied at 170 kg N ha^{-1} or no input) and previous crop management (PCM_{org} and PCM_{con}) on the abundance of individual tuber protein spots was compared, a significant main effect of fertilization and/or previous crop management and/or a significant interaction between the two factors could be detected for 98 protein spots at the $P = 0.01$ level (Figure 2). Twenty one protein spots were significantly affected by previous crop management, 33 by fertilization regime, with 9 protein spots affected by both previous crop management and fertilization regime. A significant interaction between the two factors was detected for 16 protein spots (Figure 2). The associations between these protein spots and the agronomic factors included in the experiment were analyzed by RDA (see Figure 3 described below).

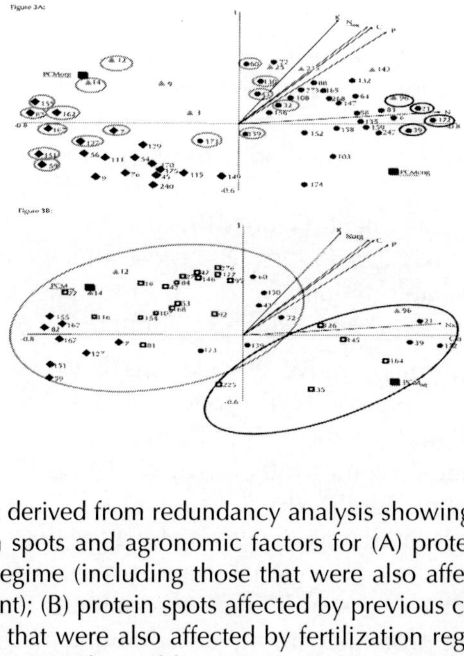

Figure 3: Bi-plot derived from redundancy analysis showing the relationship between protein spots and agronomic factors for (A) protein spots affected by fertilization regime (including those that were also affected by previous crop management); (B) protein spots affected by previous crop management (including those that were also affected by fertilization regime). Agronomic factors were: categorical variable, organic and conventional previous soil management, PCM_{org}, PCM_{con}, indicated by black squares; continuous variables, C, P, K, organic-N [N_{org}], available N [N_{av}]. Symbols indicate protein spots that had significantly increased abundance ($p < 0.01$) when no fertilizer (diamonds), high (170 kg N ha^{-1}) composted cattle manure (red triangles) or high (170 N ha^{-1}) composted chicken manure pellets (circles) was applied. The halos indicate protein spots that had significantly increased abundance ($p < 0.01$) when crops were grown in soils that were previously under either conventional crop management (black halo) or organic crop management (red halo). Transparent squares (Figure 3B only), correspond to protein spots shown by ANOVA not to be affected by fertilization regime.

When the effect of fertilizer type and input level was compared in only organic previous crop management (PCM_{org}) the abundance of 81 protein spots was significantly affected by fertilizer input type and/or levels at the $P = 0.01$ level. The associations between these protein spots and the agronomic factors included in the experiment were analyzed by RDA (see Figure 4 below).

Associations between Agronomic Factors and Tuber Protein Profiles

Only proteins that were shown to be significantly affected by fertilization and/or previous soil management (Figure 2) were included in the RDA. The biplots in Figure 3 show the relationship between agronomic factors (mineral nutrient input from organic fertilizers and previous soil management) and proteins whose abundance was shown by ANOVA to be significantly affected by fertilization (Figure 3A) or previous crop management (Figure 3B). Axis 1 explained most (33%) of variation and axis 2 a further 5%. Most additional variance was explained by available N input ($F = 7.37$, $P = 0.004$), previous crop management ($F = 6.9$, $P = 0.004$) and C-input ($F = 2.5$, $P = 0.032$).

There were strong positive associations between nutrient input levels (NH_4/NO_3-N, organic-N, P, C and K) and (a) most proteins with increased abundance in tubers fertilized with composted chicken manure pellets and (b) some proteins (25, 96, 143, 239) with increased abundance in tubers fertilized with composted cattle manure along the positive axis 1. Similar associations along the positive axis 1 were detected for previous conventional crop management (Figure 3A,B). In contrast, negative association between nutrient input levels were detected for most proteins with increased abundance in tubers from non-fertilized plants and some proteins (3, 9, 12, 14) with increased abundance in tubers fertilized with cattle manure. Proteins 3, 9, 12 and 14 were also strongly positively associated with previous organic crop management along the negative axis 1 and the positive axis 2 (Figure 3A,B).

Four protein spots that were significantly affected by previous crop management, but not fertilization regime (Figure 3B) showed negative associations between protein abundance and nutrient input levels, but positive association with previous organic crop management. Only proteins 35, 126, 145 and 164 showed the opposite trend (Figure 3B). There were strong positive associations between nutrient input levels (NH_4/NO_3-N, organic-N, P, C and K) and most proteins with increased abundance in tubers fertilized with composted chicken manure pellets and some proteins (25, 96, 143, 239) with increased abundance in tubers fertilized with composted cattle manure along the positive axis 1 (Figure 3A). Similar associations along the positive axis 1 were detected for previous conventional crop management. In contrast, negative association between nutrient input levels were detected for most proteins with increased abundance in tubers from non-fertilized plants and some proteins (3, 9, 12, 14) with increased abundance in tubers fertilized with composted cattle manure. Proteins 3, 9, 12 and 14 were also strongly positively associated with previous organic crop management along both the negative axis 1 and the positive axis 2 (Figure 3A).

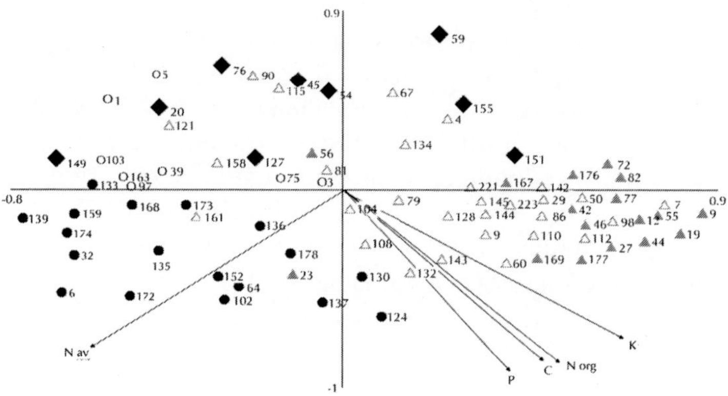

Figure 4: Bi-plot derived from redundancy analysis showing the relationship between protein spots and agronomic factors (categorical variable, organic and conventional previous soil management, PCM_{org}, PCM_{con}, indicated by black squares; continuous variables, C, P, K, organic-N [N_{org}], available N [N_{av}]). Symbols indicate protein spots that had significantly increased abundance ($p < 0.01$) when no fertilizer (diamonds), low composted cattle manure (transparent red triangles, 85 kg N ha^{-1}), high composted cattle manure (solid red triangle, 170 kg N ha^{-1}), low composted chicken manure pellets (transparent circles, 85 kg N ha^{-1}), or high chicken manure pellets (solid circles, 170 kg N ha^{-1}) was applied.

Associations between Fertilization (Types and Input Levels) and Tuber Protein Profiles

Only proteins that were shown to be significantly ($p < 0.01$) affected by fertilization in soils previously managed to organic farming standards were included in the RDA. The biplot in Figure 4 shows the relationship between nutrient input from organic fertilizers and proteins with significantly increased abundance with different fertilizer treatments (input types and levels). Axis 1 explained most (30.5%) of variation and axis 2 a further 3%. Most additional variance was explained by available N ($F = 6.82$, $P = 0.008$) and total K inputs ($F = 4.67$, $P = 0.012$).

There were strong positive associations between available-N (NH_4/NO_3-N) and most proteins with increased abundance in tubers fertilized with composted chicken manure pellets at the higher (175 kg ha^{-1}) input level along both the negative axis 1 and 2. Only 5 proteins (3, 75, 97, 103 and 163) showed significantly increased abundance at the lower (85 kg N ha^{-1}) composted chicken manure pellets input level, and these were positively associated with available N input along the negative axis 1 only. In contrast, most proteins with increased abundance in tubers fertilized with composted cattle manure at both the higher and lower input level were positively associated with K, C, P and organic N inputs along the positive axis 1 (Figure 4). All proteins with significantly increased abundance in tubers from non-fertilized control plants showed were negatively associated with nutrient inputs along axis 2.

Protein Inference by Gel Matching

Of the protein spots whose abundance was significantly ($p < 0.01$) affected by fertilization and / or previous crop management (PCM) treatments, the function of 47 was inferred by gel matching using 3 published potato tuber proteome maps [14,31,32] and the reference proteome map generated in this study (Figure 5). Proteins fell into six distinct functional groups: ATP synthesis/binding (2 protein spots), disease/defense/stress response (16 protein spots), energy production-glycolysis (11 protein spots), metabolism (3 protein spots), protein transport/destination/storage (11 protein spots), and protein breakdown (proteolysis/catabolism, 2 protein spots) (Table 4).

Some trends between the proteins inferred function and treatment that the protein spot was most highly abundant in were observed. Of the 16 disease/defense/stress response proteins, 8 had highest abundance in the composted chicken manure pellet (high rate) fertilization regime, whilst 3 had highest abundance in the control and 1 protein spot had highest abundance in the composted cattle manure (high rate) fertilization regime. Five protein spots were affected by PCM; 3 with higher abundance in PCM_{con} and 2 with higher abundance in PCM_{org}. An additional 3 defense related protein spots were affected by the fertilization × PCM interaction effect. Of the 11 protein spots with inferred roles in glycolysis, 6 had higher abundance in tubers from the composted cattle manure fertilizer regime; 3 protein spots in the low rate and 3 in the high rate of fertilizer input. Two glycolysis proteins had higher abundance in the tubers from the composted chicken manure pellets (high) fertilizer regime. Six protein spots were also affected by PCM: 4 with higher abundance in PCM_{org} and 2 with higher abundance in PCM_{con}.

Table 4: Abundance of protein spots (inferred by gel matching) that was significantly affected by fertilization (including protein spots affected by PCM (A), PCM (but not fertilization (B), a fertilization × PCM interaction effect (C)

Protein spot	Protein inference by gel matching	Function	Treatment with greatest protein spot abundance [1]	Reference protein spot [2]
A. Protein spots significantly affected by fertilization treatments (and PCM)				
176	Actin-54 (*N. tabacum*) Actin	ATP binding	Cattle (high)	4602 [31]
				38 [32]
4	14-3-3 protein	Disease/defence	Cattle (high)	1315 [31]
23	Enolase	Energy-Glycolysis	Cattle (high)	4707 [31],
				66 [32]
169	Enolase	Energy-Glycolysis	Cattle (high)	64 [32],
				5711 [31]
2	Putative nascent polypeptide associated complex *a*-chain/ expressed protein	Protein destination and storage	Cattle (high)	40 [32]

55	ATP binding/hydrolase/ nucleosidetriphosphatase/ nucleotide binding (*A. thaliana*); Protein of AAA family (*C. annuum*)	Protein synthesis, storage and turnover	Cattle (high)	3827 [14]
177	Enolase	Energy-Glycolysis	Cattle (high), PCM_{org}	14 [32]
158	Malate dehydrogenase, cytosolic; Glyceraldehyde 3-phosphate dehydrogenase	Energy-Glycolysis	Cattle (low)	7 [32]
98	Enolase-like	Energy-Glycolysis	Cattle (low)	4710 [31]
165	Enolase	Energy-Glycolysis	Cattle (low)	20 [32] 4710 [31]
60	UTP-Glc-1-*P* uridylyltransferase	Metabolism	Cattle (low), PCM_{org}	45 [32]
6	Kunitz-type enzyme inhibitor/ S9C11	Disease/defence	Chicken (high)	17 [32], 1201 [31]
156	Ascorbate peroxidase	Disease/defence	Chicken (high)	42 [32], 6301 [31]
171	Proteasome subunit	Disease/defence	Chicken (high)	5209 [31]
91	Fructokinase	Metabolism	Chicken (high)	31 [32], 4423 [31]
147	Chaperonin 21 precursor	Protein destination and storage	Chicken (high)	76 [32]
163	Cys proteinase precursor	Protein destination and storage	Chicken (high)	7205 [31] 34 [32]
18	EST (similar to small heat shock proteins)	Stress response	Chicken (high)	4208 [31]
152	Pathogenesis related protein 10	Stress response	Chicken (high)	6101 [31]
178	Hsp20.1 protein (*L. peruvianum*)	Stress response	Chicken (high)	4102 [32]
17	Ascorbate peroxidase	Disease/defence	Chicken (high), PCM_{con}	6304 [31]
21	Ascorbate peroxidase	Disease/defence	Chicken (high), PCM_{con}	5308 [31], 67 [32]
39	Triosephosphate isomerase, cytosolic isoform (*S. chacoense*)	Energy-Glycolysis	Chicken (high), PCM_{con}	6206 [31] 6 [32]

173	Glyceraldehyde 3-phosphate dehydrogenase	Energy-Glycolysis	Chicken (high), PCM_{org}	7506 [31]
				7511 [14]
43	UTP-Glc-1-*P* uridylyltransferase	Metabolism	Chicken (high), PCM_{org}	37 [32]
64	ATP synthase *b*-chain precursor, mitochondrial	ATP synthesis	Chicken (low)	36 [32]
76	Patatin precursor	Protein destination and storage	Control	2402 [14]
149	Patatin precursor	Protein destination and storage	Control	2603 [31]
170	Patatin protein 03	Protein destination and storage	Control	4504 [31]
	Patatin			29 [32]
45	PRCI (*N. tabacum*) Proteasome subunit alpha type-6	ubiquitin-dependent protein catabolic process	Control	7209 [31]
20	EST (peptide sequences LGSHFVSENQDVSIK VAYSIVGPTHSPLR FSTSSSSTK YETGRPHSYK YETGRPHSYKLR IEKYETGRPHSYKLR)	Unclassified	Control	60 [32]
175	Patatin precursor;	Disease/defence	Control	2610 [31]
	Patatin protein	Lipid degradation and metabolism,		
59	Superoxide dismutase [Cu-Zn]	Disease/defence	Control, PCM_{org}	5 [32]
167	Patatin	Protein destination and storage	Control, PCM_{org}	26 [32]
	Patatin protein 03			3506 [31]
155	Hsp19.9 protein (*L. peruvianum*)	Stress response	Control, PCM_{org}	5108 [31]
B. Protein spots significantly affected by PCM (not fertilization)				
225	Kunitz-type enzyme inhibitor S0C11	Disease/defence	PCM_{con}	3 [32]
164	Glyceraldehyde 3-phosphate dehydrogenase	Energy-Glycolysis	PCM_{con}	8403 [14]
				8508 [31]
				1 [32]
276	Phosphoglycerate mutase	Energy-Glycolysis	PCM_{org}	4809 [31]
146	Phosphoglycerate mutase	Energy-Glycolysis	PCM_{org}	4813 [31]
C. Protein spots significantly affected by fertilization × PCM interaction effect				

1	Transporter	Protein transport	PCMcon: cattle (high)	3209 [31]
			PCM_{org}: chicken (low)	
24	Putative proteasome 20S beta1 subunit (*B. napus*)	Proteolysis	PCM_{con}: cattle (high)	7202 [31]
			PCM_{org}: control	
174	Pathogenesis-related protein 10 (*S. virginianum*);	Disease/defence	PCM_{con}: control	3107 [31]
			PCM_{org}: chicken (high)	
82	Patatin; Phosphoenolpyruvate carboxykinase	Protein destination and storage;	PCM_{con}: control	28 [32],
		signal transduction	PCM_{org}: cattle (high)	4501 [31]
162	Patatin;	Protein destination and storage;	PCM_{con}: control	27 [32]
	Phosphoenolpyruvate carboxykinase	Signal transduction		
	Patatin protein 07	Protein destination and storage	PCM_{org}: cattle (high)	3509 [31]
114	Aspartic proteinase inhibitor 11 (AllergenSola t 2)	Stress response	PCM_{con}: control	4105 [31]
			PCM_{org}: cattle (high)	
129	Pathogenesis related protein 10	Stress response	PCM_{con}: control	5103 [31]
	Pathogenesis-related protein STH-2	Disease/defence	PCM_{org}: cattle (high)	2 [32]

Notes: [1]The category of treatment is listed according to the factor that significantly affected protein spot abundance (i.e., organic or conventional Previous Crop Management—PCMorg or PCMcon; fertilizer—control (no fertilizer), chicken [composted chicken manure pellets, low (85 kg N ha−1) or high (170 kg N ha−1)] or cattle [composted cattle manure low (85 kg N ha−1) or high (170 kg N ha−1)]. Where fertilizer and PCM were both significant, the category of fertilizer is listed first. In cases where there was a significant fertilizer × PCM interaction, the category of fertilizer is listed for each PCM; 2 The protein spot number and published reference (in brackets) used for protein inference by gel matching.

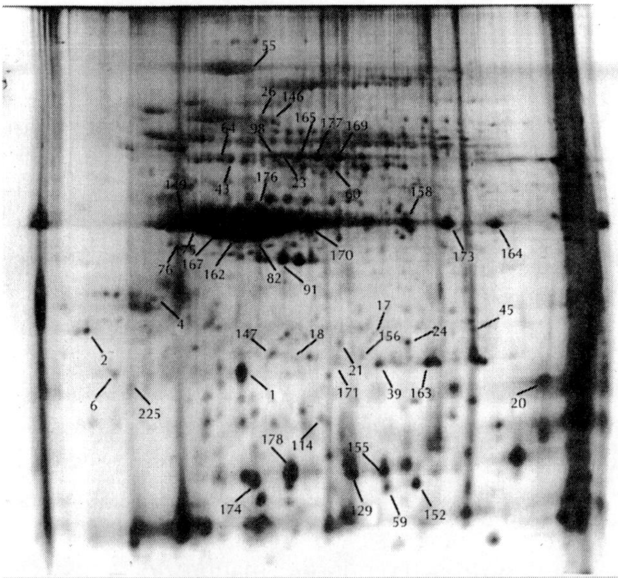

Figure 5: 2D reference gel image of potato tuber proteome used for inference of protein function by gel matching to published potato tuber proteome maps.

DISCUSSION

Effects of Previous Crop Management, Fertilization Regime, and Watering Supply on Potato Traits

Reducing the reliance on mineral fertilizers by increased use of organic fertilizers such as manure and composted organic waste (including sewage) is thought to be an important strategy to improve the sustainability of crop production and future food security (1,2,3). However, replacing mineral fertilizers with composted cattle manure was previously shown to significantly reduce yields and affect gene and protein expression pattern in potato, even when the maximum input level permitted under current Environmental legislation (170 kg ha^{-1} annum^{-1}) is used [13,14,33]. This is thought to be mainly due to insufficient supply of nitrogen from organic fertilizers, such as composted manure, which contain very low concentrations of water soluble, readily plant available forms of nitrogen (NO_3^-, NH_4^+). However, the omission of chemosynthetic pesticides in organic potato production also contributes to the yield differential between organic and conventional crops (18). Results reported here show that the use of composted chicken manure pellets (which have a higher concentration of readily plant available nitrogen) will significantly increase potato yields compared to cattle manure compost at the same total N-input level. The use of composted chicken manure pellets and/or composts supplemented with organic fertilizers with high plant available-N content (e.g., manure slurry or vinasse, a by-product of the sugar industry) may therefore significantly increase yields in organic and other sustainable farming systems which aim to reduce mineral fertilizer inputs.

Nitrogen supply from composted manure to crops is reliant on microbial mineralization processes in soil and the total amount of nitrogen becoming available from fertilizer applied prior to the planting is known to be lower when organic rather than mineral fertilizers are used (12,18). Soil microbial activity (and associated mineralization capacity) was shown to increase in soils which receive regular organic fertilizers over long periods of time and both soil microbial activity and plant growth are reduced by suboptimal irrigation/water availability in soils (12, 14, 16). Potential strategies to improve crop yields from organic fertilizer inputs therefore include (a) the use of organic fertilizers with a higher content of available forms of N (e.g. chicken manure), (b) optimization of water supply and (c) more frequent use of organic fertilizer inputs (to increase mineralization capacity in soils). The higher levels of chlorophyll in leaves of plants fertilized with composted chicken manure pellets and RDA results observed in this study suggest that the use of composted chicken manure pellets (which contain 20 times higher concentrations of readily plant available N in form of NH_4^+) increased crop yield mainly *via* increasing N-supply to crops. N-supply is an important factor affecting photosynthesis with N deficiency reducing chlorophyll synthesis and growth of photosynthetic plant organs (stems and leaves) and resulting in early senescence [34]. Composted chicken manure pellets may also have stimulated mineralization, since previous studies showed poultry manure (but not cattle manure) inputs increases populations of bacterial feeding soil nematodes, which indicates increased microbiological activity and nutrient cycling [35].

In this study, previous organic crop management (which involved regular composted manure inputs over a 4 year period) reduced potato yields, when compared to conventional previous soil management (which involved regular mineral fertilizer inputs). This was unexpected, because previous comparative studies reported that replacing mineral fertilizer with manure inputs will result in an increase in soil microbiological activity and diversity after only two years [18,19]. Soil biological activity/mineralization capacity was not assessed in this study, but results suggests that previous conventional crop management/ mineral fertilization resulted in higher residual soil nutrient levels and that this had a greater effect on nutrient supply and crop yields than possible effects of regular organic fertilizer inputs on soil biological activity/mineralization capacity [11]. Further studies should therefore focus on quantifying the effect of previous crop management practices on microbial activity/mineralization capacity, soil nitrogen cycling and crop performance.

As expected, restricted watering resulted in lower potato tuber yield. However, when compared at the same fertilization regime, restricted watering resulted in a higher leaf chlorophyll content (which is closely correlated to N-availability), suggesting that restricted water supply restricted potato yield primarily *via* direct effects on plant growth rather than indirectly *via* reducing soil mineralization processes and N-supply [36,37,38]. Water stress is known to result in higher stomatal resistance in plant leaves as a way to reduce water loss, but this limits CO_2 diffusion into the sub-stomatal cavities, thus limiting photosynthesis [34].

Effects of Previous Crop Management and Fertilization Regime on the Potato Tuber Proteome

Previous studies showed that a switch from mineral fertilizer to composted cattle manure inputs results in significant differences in protein expression pattern, with more than 20% of protein spots being differentially expressed. In contrast, the use of different crop protection regimes (use or non-use of chemosynthetic pesticides) can have a very limited impact on the potato proteome [14]. Here we demonstrated that both previous crop management (organic *vs.* conventional) and contrasting organic fertilizer input types (composted chicken manure pellets and composted cattle manure) can result in significant changes in protein expression. Results from the RDA suggested that a main driver for contrasting protein expression was the difference in available N supplied by different fertilization treatments. The abundance of some proteins was affected by both previous crop management (PCM) and fertilization regimes indicating that the impact of PCM on these proteins was linked to the differences in nutrient availability associated with PCM. Protein inference by gel matching showed that the up-regulation of functional groups (e.g., plant defense, glycolysis) was not restricted to a single plant treatment. Rather, different proteins within the same functional groups were differentially up-regulated due to contrasting PCM/fertilization treatments. For example, different stages of the glycolysis pathway were up-regulated in the different treatments. Since proteomics is a "snapshot" analysis (*i.e.*, it only shows the differences in plant responses at a given moment in time) this data indicates that there are differences in the rate or timing of glycolysis between the treatments leading to the observed differences in abundance of proteins involved in the different stages of glycolysis. For other functional groups (e.g., defence/stress and protein destination), the different proteins are not linked as part of a single pathway. Therefore, this indicates that there is a specificity in the function that each protein performs (and therefore the molecular activity occurring in the tubers), according to the plant environment (treatment) that it is up-regulated in. Lehesranta *et al.* [14] previously found that the use of organic (composted cattle manure) fertilizer led to a higher stress response in potato tubers compared to the use of mineral fertilizer. In this study, we found that contrasting types and levels of organic fertilizer can also affect the stress response in potato tubers. A higher number of protein spots with an inferred role in stress response had higher abundance in tubers of composted chicken manure fertilized plants. In studies such as this, where mass spectrometry based protein identification was not feasible, gel matching can be used to provide interesting inferences of what the protein function could be [30]. Results from gel matching should be treated with caution due to uncertainty that can arise when comparing the spatial pattern of protein spots on 2D gels, especially across different studies and/or plant genotypes/varieties/cultivars. Care should be taken to only match gels that were run using the same 2DE conditions, as this will drastically affect the spatial pattern of protein spots. Whenever gel matching is used, attention should be given to the criteria employed to accept/reject inferred protein function, as cases can arise where e.g. variability between gels could cause errors leading to increased likelihood of false inference.

The results presented here demonstrate our ability to quantify a molecular response to contrasting types and rates of organic fertilizers that is associated with genes involved in nutrient use efficiency. This is an important finding that indicates the importance of optimizing the constitution of organic fertilizers to reduce yield loss when they are used instead of mineral fertilizers. Further study will be required to extrapolate to other potato varieties, to investigate whether genetic variation influences nutrient use efficiencies in contrasting types of organic fertilizers. The use of proteomics enables linkage between genomic pathways controlling crop responses to contrasting fertilization (nutrient use efficiency and yield) and the respective encoding genes [14,20,21,39], which in future studies may be used complementary to transcriptomics analysis [13,33] for understanding gene expression pathways. This will facilitate the development of functional molecular markers to accelerate future crop breeding programs [22] that seek to improve crop yields and quality from organic fertilizers based on animal and green manures and recycled domestic organic waste (including sewage). Such strategies may be crucial to improve the sustainability, resource use efficiency and environmental impacts of crop production [1,3,4,5,6].

ACKNOWLEDGMENTS

The authors gratefully acknowledge funding from the European Community (EU Fifth Framework project QLK5-CT-2000-01065: Blight-MOP—development of a systems approach for the management of late blight in EU organic potato production) and technical staff for their assistance.

SUPPLEMENTARY FILES

Supplementary File 1:

Table S1: Effects of previous crop management (PCM), fertilization regime (FR), and watering regime (WR) on leaf SPAD measurements (units), 41 and 81 days after emergence

6

Factor	**41 days**	**81 days**
Means (± SE)		
WR		
Restricted	40.3 ± 0.4	35.9 ± 0.7
Optimized	37.6 ± 0.3	29.4 ± 0.5
PCM		
Conventional soil	39.3 ± 0.4	32.6 ± 0.8
Organic soil	38.6 ± 0.3	32.7 ± 0.6
FR		
Control 0 N ha^{-1}	37.2 ± 0.5 c	30.1 ± 0.9 c
Composted cattle manure, 85 kg N ha^{-1}	38.0 ± 0.6 c	31.3 ± 1.0 c
Composted chicken manure pellets, 85 kg N ha^{-1}	39.8 ± 0.5 b	33.6 ± 1.2 b
Composted cattle manure, 170 kg N ha^{-1}	37.9 ± 0.6 c	31.4 ± 1.1 c
Composted chicken manure pellets, 170 kg N ha^{-1}	41.8 ± 0.6 a	37.0 ± 0.9 a
ANOVA		
Main effects		
WR	0.0001	0.0009
PCM	0.0313	ns
FR	<0.0001	<0.0001
Interaction effects		
WR × PCM	0.0427	0.0002
WR × FR	0.0908	ns
PCM × FR	ns	ns
WR × PCM × FR	ns	ns

REFERENCES

1. Tilman, D.; Cassman, K.G.; Matson, P.A.; Naylor, R.; Polasky, S. Agricultural sustainability and intensive production practices. *Nature* 2002, *418*, 671–677.
2. Flynn, H.C.; Smith, P. *Greenhouse Gas Budgets of Crop Production—Current and Likely Future Trends*; International Fertilizer Industry Association: Paris, France, 2010.
3. Robertson, G.P.; Vitousek, P.M. Nitrogen in agriculture: Balancing the cost of an essential resource. *Ann. Rev. Environ. Resour.* 2009, *34*, 97–125.
4. Cordell, D.; Dangert, J.-O.; White, S. The story of phosphorus: Global food security and food for thought. *Glob. Environ. Change* 2009, *19*, 292–305.
5. Fantel, R.J.; Peterson, G.R.; Stowasser, W.F. The worldwide availability of phosphate rock. *Nat. Resour. Forum* 1985, *9*, 5–24.
6. Trehan, S.P.; Sharma, R.C. Differences in phosphorus use efficiency in potato genotypes. *Adv. Hortic. Sci.* 2005, *19*, 13–20.
7. Hepperly, P.; Lotter, D.; Ziegler, C.; Seidel, R.; Reider, C. Compost, manure and synthetic fertilizer influences crop yields, soil properties, nitrate leaching and crop nutrient content. *Compost Sci. Util.* 2009, *17*, 117–126.
8. Herencia, J.F.; Ruiz-Porras, J.C.; Melero, S.; Garcia-Galavis, P.A.; Morillo, E.; Maqueda, C. Comparison between organic and mineral fertilization for soil fertility levels, crop macronutrient concentrations, and yield. *Agron. J.* 2007, *99*, 973–983.
9. Warman, P.R.; Burnham, J.C.; Eaton, L.J. Effects of repeated applications of municipal solid waste compost and fertilizers to three lowbush blueberry fields. *Sci. Hortic.* 2009, *122*, 393–398.
10. Eyre, M.D.; Sanderson, R.A.; Shotton, P.N.; Leifert, C. Investigating the effects of crop type, fertility management and crop protection on the activity of beneficial invertebrates in an extensive farm management comparison trial. *Ann. Appl. Biol.* 2009, *155*, 267–276.
11. Bulluck, L.R.; Brosius, M.; Evanylo, G.K.; Ristaino, J.B. Organic and synthetic fertility amendments influence soil microbial,

physical and chemical properties on organic and conventional farms. *Appl. Soil Ecol.* 2002, *19*, 147–160.

12. The Council Of The European Communities. Council Directive 91/676/EEC of 12 December 1991 concerning the protection of waters against pollution caused by nitrates from agricultural sources. *Off. J. Eur. Union* 1991, *91/676/EEC*, 1–8.
13. Van Dijk, J.P.; Cankar, K.; Hendriksen, P.J.M.; Beenen, H.G.; Zhu, M.; Scheffer, S.; Shepherd, L.V.T.; Stewart, D.; Davies, H.V.; Leifert, C.; Wilkockson, S.J.; Gruden, K.; Kok, E.J. The assessment of differences in the transcriptomes of organically and conventionally grown potato tubers. *J. Agric. Food Chem.* 2012, *60*, 2090–2101.
14. Lehesranta, S.J.; Koistinen, K.M.; Massat, N.; Davies, H.V.; Shepherd, L.V.T.; McNicol, J.W.; Cakmak, I.; Cooper, J.; Luck, L.; Karenlampi, S.O.; Leifert, C. Effects of agricultural production systems and their components on protein profiles of potato tubers. *Proteomics* 2007, *7*, 597–604.
15. Mäder, P.; Fliessbach, A.; Dubois, D.; Gunst, L.; Fried, P.; Niggli, U. Soil fertility and biodiversity in organic farming. *Science* 2002, *296*, 1694–1697.
16. Chang, E.H.; Chung, R.S.; Wang, F.N. Effect of different types of organic fertilizers on the chemical properties and enzymatic activities of an Oxisol under intensive cultivation of vegetables for 4 years. *Soil Sci. Plant Nutr.* 2008, *54*, 587–599.
17. Ceylan, S.; Mordogan, N.; Akdemir, H.; Cakici, H. Effect of organic fertilizers on some agronomic and chemical properties of potato (*Solanum tuberosum* L.). *Asian J. Chem.* 2006, *18*, 1223–1230.
18. Gunapala, N.; Scow, K.M. Dynamics of soil microbial biomass and activity in conventional and organic farming systems. *Soil Biol. Biochem.* 1998, *30*, 805–816.
19. Shannon, D.; Sen, A.M.; Johnson, D.B. A comparative study of the microbiology of soils managed under organic and conventional regimes. *Soil Use Manag.* 2002, *18*, 274–283.
20. Rempelos, L.; Cooper, J.; Wilcockson, S.; Eyre, M.; Shotton, P.; Volakakis, N.; Orr, C.H.; Leifert, C.; Gatehouse, A.M.R.; Tétard-Jones, C. Quantitative proteomics to study the response of potato to contrasting fertilisation regimes. *Mol. Breed.* 2012. in press.

21. Tétard-Jones, C.; Shotton, P.N.; Rempelos, L.; Cooper, J.; Eyre, M.; Orr, C.H.; Leifert, C.; Gatehouse, A.M.R. Quantitative proteomics to study the response of wheat to contrasting fertilisation regimes. *Mol. Breed.* 2012. in press.

22. Varshney, R.K.; Graner, A.; Sorrells, M.E. Genomics-assisted breeding for crop improvement. *Trends Plant Sci.* 2005, *10*, 621–630.

23. Hack, H.; Gall, H.; Klemke, T.H.; Klose, R.; Meier, U.; Stauss, R.; Witzen-Berger, A. Phänologische Entwicklungsstadien der Kartoffel (*Solanum tuberosum* L). Codirung und Beschreibung nach der erweiterten BBCH-Skala mit Abbildungen. *Nachrichtenbl. Deut. Pflanzenschutzd.* 1993, *45*, 11–19.

24. Hurkman, W.J.; Tanaka, C.K. Solubilization of plant membrane proteins for analysis by two-dimensional gel electrophoresis. *Plant Physiol.* 1986, *81*, 802–806.

25. Healthcare, G.E. *2-D Electrophoresis Principles and Methods*; General Electric Company: Uppsala, Sweden, 2010.

26. Pinheiro, J.C.; Bates, D.M. *Mixed-Effects Models in S and S-PLUS*; Springer: New York, NY, USA, 2000.

27. Crawley, M.J. *The R Book*; John Wiley and Son: Chichester, UK, 2007.

28. Team, R.D.C. *R: A language and environment for statistical computing. R Foundation for Statistical Computing*; R Foundation for Statistical Computing: Vienna, Austria, 2009.

29. Ter Braak, C.J.F.; Šmilauer, P. *CANOCO Reference Manual and User's Guide to Canoco for Windows: Software for Canonical Community Ordination*, 4th ed.; Centre for Biometry: Wageningen, The Netherlands, 1998.

30. Chibani, K.; Ali-Rachedi, S.; Job, C.; Job, D.; Jullien, M.; Grappin, P. Proteomic analysis of seed dormancy in arabidopsis. *Plant Physiol.* 2006, *142*, 1493–1510.

31. Lehesranta, S.J.; Davies, H.V.; Shepherd, L.V.T.; Koistinen, K.M.; Massat, N.; Nunan, N.; McNicol, J.W.; Kärenlampi, S.O. Proteomic analysis of the potato tuber life cycle. *Proteomics* 2006, *6*, 6042–6052.

32. Lehesranta, S.J.; Davies, H.V.; Shepherd, L.V.T.; Nunan, N.; McNicol, J.W.; Auriola, S.; Koistinen, K.M.; Suomalainen, S.;

Kokko, H.I.; Kärenlampi, S.O. Comparison of tuber proteomes of potato varieties, landraces, and genetically modified lines. *Plant Physiol.* 2005, *138*, 1690–1699.

33. Van Dijk, J.P.; Cankar, K.; Scheffer, S.J.; Beenen, H.G.; Shepherd, L.V.T.; Stewart, D.; Davies, H.V.; Wilkockson, S.J.; Lelfert, C.; Gruden, K.; Kok, E.J. Transcriptome analysis of potato tubers-effects of different agricultural practices. *J. Agric. Food Chem.* 2009, *57*, 1612–1623.
34. Olesinski, A.A.; Wolf, S.; Rudich, J.; Marani, A. The effect of nitrogen fertilization and irrigation frequency on photosynthesis of potatoes (*Solarium tuberosum*). *Ann. Bot.* 1989, *64*, 651–657.
35. Griffiths, B.S.; Ritz, K.; Wheatley, R.E. Nematodes as indicators of enhanced microbiological activity in a Scottish organic farming system. *Soil Use Manag.* 1994, *10*, 20–24.
36. Vos, J.; Groenwold, J. Water relations of potato leaves, I. Diurnal changes, gradients in the canopy, and effects of leaf-insertion number, cultivar and drought. *Ann. Bot.* 1988, *62*, 363–371.
37. Songsri, P.; Jogloy, S.; Holbrook, C.C.; Kesmala, T.; Vorasoot, N.; Akkasaeng, C.; Patanothai, A. Association of root, specific leaf area and SPAD chlorophyll meter reading to water use efficiency of peanut under different available soil water. *Agric. Water Manag.* 2009, *96*, 790–798.
38. Barraclough, P.B.; Kyte, J. Effect of water stress on chlorophyll meter readings in winter wheat. In *Plant Nutrition—Food Security and Sustainability of Agro-ecosystems*; Horst, W.J., Schenk, M.K., Burkert, A., Claassen, N., Flessa, H., Frommer, W.B., Goldbach, H., Olfs, H.W., Romheld, V., Sattelmacher, B., Schmidhalter, U., Schubert, S., Wiren, N., Wittenmayer, L., Eds.; Springer: Dordrecht, The Netherlands, 2001; Volume 92, pp. 722–723. Developments in Plant and Soil Sciences.
39. Nawrocki, A.; Thorup-Kristensen, K.; Jensen, O.N. Quantitative proteomics by 2DE and MALDI MS/MS uncover the effects of organic and conventional cropping methods on vegetable products. *J. Proteomics* 2011, *74*, 2810–2825.

Chapter 7

Data Mining Methods for Omics and Knowledge of Crude Medicinal Plants toward Big Data Biology

Farit M. Afendi[a, b], Naoaki Ono[a], Yukiko Nakamura[a], Kensuke Nakamura[d], Latifah K. Darusman[c], Nelson Kibinge[a], Aki Hirai Morita[a], Ken Tanaka[e], Hisayuki Horai[f], Md. Altaf-Ul-Amin[a], and Shigehiko Kanaya[a,*]

[a]Graduate School of Information Science, Nara Institute of Science and Technology, Nara 630-0101, Ikoma, Japan

[b]Department of Statistics, Bogor Agricultural University, Jln. Meranti, Kampus IPB Darmaga, Bogor 16680, Indonesia

[c]Biopharmaca Research Center, Bogor Agricultural University, Kampas IPB Taman Kencana, Jln. Taman Kencana No. 3 Bogor 16151, Indonesia

[d]Maebashi Institute of technology, 450-1 Kamisadori, Maebashi-shi, Gunma, 371-0816 Japan

[e]Department of Medicinal Resources, Institute of Natural Medicine, University of Toyama, 2630 Toyama, 930-0194, Japan

[f]Department of Electronic and Computer Engineering, Ibaraki National College of Technology, 866 Nakane, Hitachinaka, Ibaraki 312-8508, Japan

ABSTRACT

Molecular biological data has rapidly increased with the recent progress of the Omics fields, e.g., genomics, transcriptomics, proteomics and metabolomics that necessitates the development of databases and methods for efficient storage, retrieval, integration and analysis of massive data. The present study reviews the usage of KNApSAcK Family DB in metabolomics and related area, discusses several statistical methods for handling multivariate data and shows their application on Indonesian blended herbal medicines (Jamu) as a case study. Exploration using Biplot reveals many plants are rarely utilized while some plants are highly utilized toward specific efficacy. Furthermore, the ingredients of Jamu formulas are modeled using Partial Least Squares Discriminant Analysis (PLS-DA) in order to predict their efficacy. The plants used in each Jamu medicine served as the predictors, whereas the efficacy of each Jamu provided the responses. This model produces 71.6% correct classification in predicting efficacy. Permutation test then is used to determine plants that serve as main ingredients in Jamu formula by evaluating the significance of the PLS-DA coefficients. Next, in order to explain the role of plants that serve as main ingredients in Jamu medicines, information of pharmacological activity of the plants is added to the predictor block. Then N-PLS-DA model, multiway version of PLS-DA, is utilized to handle the three-dimensional array of the predictor block. The resulting N-PLS-DA model reveals that the effects of some pharmacological activities are specific for certain efficacy and the other activities are diverse toward many efficacies. Mathematical modeling introduced in the present study can be utilized in global analysis of big data targeting to reveal the underlying biology.

INTRODUCTION

Data-intensive sciences have progressed in modern astronomy [1], biology [2, 3, 4, 5, 6, 7 and 8], computational materials science [9], ecology [10 and 11] and social science [12] because open-access data has increased drastically. Data-intensive or -driven discovery in biology requires a large open pool of data across the full breadth of the life sciences and the access to the pool will invite "New" logic, strategies and tools to discover new trends, associations, discontinuities, and exceptions that reveal aspects of the underlying biology [2, 5 and 6]. Big data biology, which is a discipline of data-intensive science, was proposed based on the rapid increasing of omics data produced by genomics, transcriptomics, proteomics and metabolomics [2, 3, 4, 5, 6, 7 and 8]. This situation is also a feature of the ethnomedicinal survey and the number of medicinal plants is estimated to be 40,000 to 70,000 around the world [13] and many countries utilize these plants as blended herbal medicines, e.g., China (traditional Chinese medicine), Japan (Kampo medicine), India (Ayruveda, Siddha and Unani) and Indonesia (Jamu). Blended herbal medicines as well as single herb medicines include a large number of constituent substances which exert effects on human physiology through a variety of biological pathways. To comprehensively understand the medicinal usage of plants based upon traditional and modern knowledge, we add to KNApSAcK Family database systems the selected herbal ingredients i.e., the formulas of Kampo and Jamu, omics information in plants and humans, and physiological activities in humans [14, 15 and 16]. These information need to be connected in a way that enables scientists to make predictions based on general principles.

In this mini-review, we discuss the usage of KNApSAcK Family DB in metabolomics, explain mining techniques such as principal component analysis (PCA), partial least square regression (PLSR) and multiway model, and show their application on Indonesian blended herbal medicines (Jamu) as a case study.

KNAPSACK FAMILY DATABASE

Omics biology, like most scientific disciplines, is in an era of accelerated increase of data, so called big data biology [2, 3, 4, 5, 6, 7 and 8]. Large-scale sequencing centers, high-throughput analytical facilities and individual laboratories produce vast amounts of data such as nucleotide and protein sequences, gene expression measurements, protein and genetic interactions, mass spectra of metabolites and phenotype studies. The goal of investigating the interactions between medicinal/edible plants and humans is to comprehensively understand the molecular mechanism of medicinal plants on human physiology based on current and traditional knowledge. Optimization of blended herbal formulas should be developing using information derived from plant and human omics. To reach this goal we need to develop databases based on the platform shown in Fig. 1A. KNApSAcK family DBs have been developed for this purpose [14,15 and 16]. Relations among individual DBs are illustrated in Fig. 1A and main page of KNApSAcK Family DB is shown in Fig. 1B.

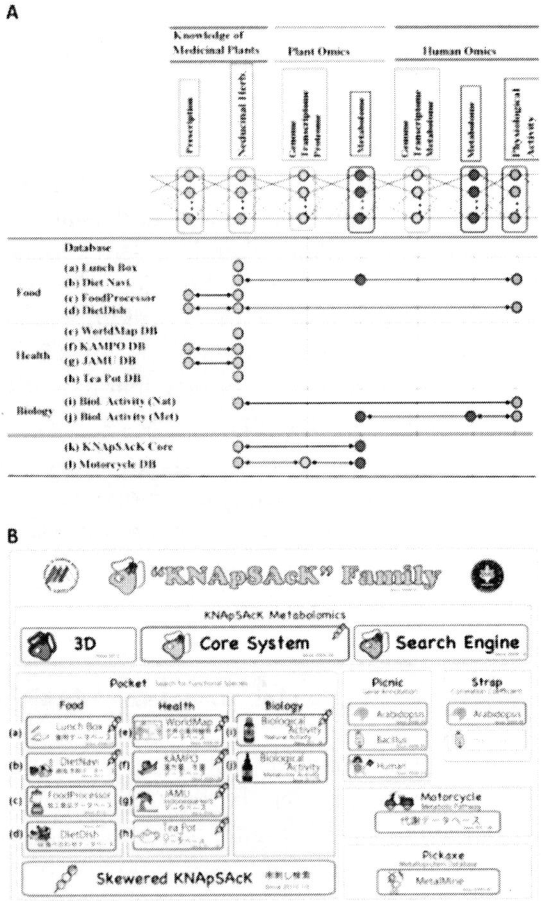

Figure 1: Integrated platform of knowledge of medicinal plants and plant and human –omics and KNApSacK Family databases. (A) The relations of attributes among individual DBs. (B) Main window of KNApSAcK Family DB, indexes from a to i in panel A correspond to those in panel B.

Four DBs (Lunch Box DB, DietNavi DB, Food Processor DB and DietDish DB, a-d in Fig. 1) are about Food & Health related with Japanese foods and ingredients explained in Japanese language because initially we developed them targeting the Japanese people, but we are planning to translate them into English as early as possible. Lunch Box DB comprises information on 800 edible species which include the species introduced to Japan from outside or originally grown in Japan, general information of the crops and the effect of them on human health.

Noncommunicable diseases such as heart disease, metabolic disease, cancer and respiratory disease, which superseded the infectious diseases because of the development and widespread distribution of vaccines and antimicrobial drugs, account for 60% of all deaths worldwide and 80% of deaths in low- and middle-income countries [17]. Food and ingredients in sanative diet and more effective combination of foods beneficial against those noncommunicable diseases are accumulated in DietNavi and DietDish DBs, respectively (b and d in Fig. 1). FoodProcessor DB comprises 309 retortable pouch foods encompassed by 261 food ingredients produced in Japan, and connected with DietNavi and KNApSAcK core by species names of foods.

To systematize crude drugs by multifaceted view points, we have developed four DBs (WorldMap, KAMPO, JAMU and TeaPot DBs as shown in e-h of Fig. 1). The KNApSAcK WorldMap DB comprises 46,256 geographic zone-plant pair entries in 217 geographical zones except mini-states such as the Principalities of Liechtenstein, Monaco and Andorra, and the Vatican City. Prescriptions corresponding to Japanese and Indonesian herbal medicines have been accumulated in KAMPO and JAMU DBs, respectively. KAMPO DB is comprised of 1,581 primary formulas classified in to 336 formula names encompassed by 278 medicinal plants which are approved by the National health insurance authority in Japan. JAMU DB is comprised of 5,310 formulas encompassed by 550 medicinal plants and 12 anatomical regions which are approved by the National Agency of Drug and Food Control (NA-DFC) of Indonesia. Medicinal/edible plants reported in the scientific literature have been classified into geographic zones using the International Organization for Standardization (ISO3166), which defines geographic zones based on the borders between nations and small islands. Herbs are defined as any plants with leaves, seeds, and flowers used for flavoring, food, medicine, perfume and parts of such a plant as used in cooking. Those are accumulated in TeaPot DB.

Two types of biological activities, that is, activities of natural resources and metabolites to other species including human, i.e., antibiotic, anticancer and so on are accumulated in Natural Activity and Metabolite Activity DBs (Fig. 1B), respectively. The former and the latter comprised 33,703 and 6,677 entries, respectively. For extension of species-metabolite relationship DB to metabolic pathways, it is needed to design secondary metabolic pathway DB for detection of metabolic pathways based on enzyme reactions and prediction of reactions by peptide sequences. So we have developed Motoercycle DB containing 2,421 entries. The metabolomics of plants is developing rapidly [18, 19, 20, 1, 1, 2, 3, 4, 5, 6, 7 and 8 and references in Table 1], and it will be an important topic in the systems-biological studies of interactions between plants and humans, which is included in the topics of big data biology [2, 3, 4, 5, 6, 7 and 8], with the goal of achieving a holistic understanding of plant function and healthcare, including the activity of medicinal plants as well as interaction between plants and their environment [14, 15, 16, 21 and 22].

Table 1: Studies that cite KNApSAcK Core DB

Article type	The purpose of study [References]
< 2006–2008 >	
Review	Bridge between Chemistry and Biology [24], GC-MS DB [29], Metabolomics technologies [31], Functional genomics research strategy of combining transcriptome and metabolome [32], The role of MS in metabolomics [34], Mass spectrometry platforms [38], Metabolomics technologies and functional genomics platform [42], Technology and informatics [49], Atmospheric pressure ionization mass spectrometry [52]

Exp	Metabolite accumulation caused by herbicidal enzyme inhibitors [30], Assignment of UGT89C1 to a flavonol 7-O-rhamnosyltransferase [33], Light/dark regulation of metabolite activities [35], Characterization of mutants in flavonoid and phenylpropanoid biosynthetic pathways [37], Metabolism of dietary phytochemicals [39], Metabolic networks in primary and secondary pathways for achene and receptacle [40], High-resolution mass spectrometry and 13C-isotope labeling of entire metabolomes [41], Phenolic biosynthesis pathway [43], Metabolic profiling in strawberry receptacle development [44], Regulation of glucosinolate biosynthesis [46], Integrated analysis of metabolome and transcriptome [48], Protocol in metabolite fingerprints [50]
Bioinfo	Metabolome platform DrDMASS in FT-ICR-MS [25], Taxonomic diversity of flavonoids [26], MS Peak storage and processing [28], Metabolite annotation based on MS and MS2 [45], Identification of metabolites by MS and MS-tagged MS2 data [47], Metabolome platform DrDMASS in FT-ICR-MS [51]
DB	Chemical biology [27], Metabolome tools and databases [36]
< 2009 >	
Review	Integrated omics [58], MS-based technologies [59], Web-resources in MS-based metabolomics [75], Functional genomics [78]
Exp	Metabolic profiling in cold-temperature [56], Antioxidant compounds in white cabbage during winter storage [60], Hydroxylation of fatty acids by P450 proteins [62], Dietary phytochemicals and human [63], Classification of Ephedra sp. [67], Selection of metabolites [68], Matrix-assisted laser desorption/ionization mass spectrometry [69], Determination of gene function [70], Quality assessment [73 and 74], Diarylheptanoid biosynthesis [77]
Bioinfo	Annotation of metabolite information to MS [53], Tools for the annotation of High Resolution MS metabolomics data [57], Comparison of metabolite DB using rice metabolites [61], Assessment of annotation of metabolites using FDR [64], Graph representation of multiple databases [65], Peak detection based on MS/MS patterns [66], Complexity of relation between plants and metabolites [71], Metabolic pathway prediction [72], Metabolite Complexity of relation between plants and metabolites [71], Metabolic pathway prediction [72], Metabolite annotation [76]
DB	Embedded string-search commands on MediaWiki [54]
< 2010 >	
Review	MS data processing [84], Metabolomics in plant ecology and genetics [85], Identification of metabolites [87], FT-ICR-MS, Reaction representation based on van Krevelen diagram [89], Relationship among individual omics data based on multivariate analysis and DB [16], Dietary intake [90], Functional Genomics [92], Annotation of gene function based on co-response gene and identification of metabolites [95]
Exp	Metabolite composition [79], QTL of barley, against Fusarium head blight [80], Changing color of flower from dark purple to white [81], Metabolic profiling of different tissues [86], Quality assessment [94]

Bioinfo	Chemical similarity search and substructure matching of compounds [82], Multiple metabolomics platforms for different types of MS [91], MS data processing [96], Network analysis of species-metabolite relations [97]
DB	MassBank, MS DB [83], Polyphenol contents in foods [88], Binzylsioquinone alkaloids [93]
< 2011 >	
Review	Pesticide research [100], Metabolome DB [108], Traditional medicinal plants [111], Pesticide research [113]
Exp	Hepatotoxicity [55], Subcellular distribution of metabolites [99], Assessment of metabolites of barley against Fusarium head blight [102], Metabolic responses of ultraviolet-B light [103], Transport of 12-Oxo-phytodienoic acid-glutathione into vacuole [104], Demetylation of oligogalacturonides by FAPE1 leads to defense against fungus Botytis cinerea [105], Cytochrome P450, CYP81F4 [109], Imaging mass spectrometry [112]
Bioinfo	QTL informatics [98], Metabolomics in medical purpose with systems chemical biology and chemoinformatics [101], Molecular formula annotation of polar and lipophilic metabolites [107], Metabolic profiling [114]
DB	Food phytochemicals [106], Medicinal plants in Indonesia [110]
< 2012–13 >	
Review	Plant responses to abiotic stress [115], Phytoalexins [118], Plant biotechnology [119], Integrative system biology [121], Systems biology in Japanese traditional Kampo medicine [15]
Exp	Camptothecin biosynthesis [117], Herbivore (Spodoptera littoralis)-induced metabolites [120], Natural distance [122], Molecular marker [123], Metabolic changes during fruit maturation [124], Metabolites in seed kernels [125], mQLT [126], Salt and drought stress [127], Mass spectrometric imaging [130], Defence against pathogens (Penicillium digitatum) [131]
Bioinfo	Repository for metabolomics studies [128], Visualization of metabolome data [129]
DB	Metabolite annotation [116]

To facilitate access to metabolite information obtained from analytical techniques, we have developed species-metabolite relationship DB (KNApSAcK Core DB) which contains 106,418 species-metabolite relationships encompassing 21,705 species and 50,897 metabolites. Nine databases of KNApSAcK family (except DietDish) are connected with KNApSAcK Core DB to easily obtain candidates of secondary metabolites in species utilized in several purposes [23]. The KNApSAcK Core DB was utilized in very diverged purposes of metabolomics studies including identification of metabolites ('Exp' in

Table 1), construction of integrated databases ('DB'), bioinformatics and systems biology ('Bioinfo'), and cited in at least 110 papers listed in Table 1, that is, in 29 papers in the period of 2006–2008, 25 papers in the period of 2009, 20 papers in 2010, 18 papers in 2011, 18 papers in 2012–2013. In addition, it was applied in diverged species from bacteria to plants and animals, in total 28 species, that is, Angelica acutiloba [74],Arabidopsis lyrata ssp. petraea [56], Arabidopsis thaliana[25, 30, 33, 35, 37, 46, 47, 62, 70, 86, 99, 103,104, 108, 109, 121 and 122], Atriplex halimus [127], Bacillus subtilis [113], Brassica oleraceae var capitata[60], Brufelsia calycina [81], Capsicum sp. [123], Citrus sinensis [131], Curcuma longa [77], Ephedra sp. [67], Escherichia coli [51], Fragaria × ananassa [40, 43 and 44], Fragaria vesca [105], Glycine max [53],Glycyrrhiza uralensis [94], Hordeum vulgare [80 and 102], Homo sapiens [63 and 101], Jatropha curcas [124 and 125], Malx × domestica [126], Ophiorrhiza pumila [117], Oryza sativa [49 and 61], Papaver somniferum [42], Rattus norvegicus [39 and 97], Rizotania solani [79], Solanum lycopersicum [45 and 48], Solanum tuberosum [98] and Zea mays [120].

In the period of 2006–2008, many review papers ['Review' in Table 1] focused on metabolomics platforms integrated by mass-spectrometry and metabolite databases including KNApSAcK Core [29, 31, 34, 38, 42,49 and 52] and on linking chemistry with biology [24], and on metabolome researches targeting the model plant Arabidopsis thaliana [30, 33, 35 and 37]. In 2009, metabolome studies were extended to diverged species such as crops and medicinal plants [53, 60, 61, 67, 68, 73, 74 and 78] and to engineering studies such as quality assessment based on metabolomics [73 and 74]. Thus metabolomics was applied from model species to crops and medicinal herbs. In the period of 2010–2013, metabolomics was further extended to genetics such as QTL [80, 98 and 126], and to explanation of species by metabolites, i.e., ecological subjects [85] phytoalexins [119], herbivore-induced metabolites [120] and defense against pathogens [131], and to stress responses [115, 116 and 127]. In addition, metabolomics has also been tried in imaging studies [112 and 129]. Species-metabolite relation database KNApSAcK Core has been utilized in the extended fields of metabolomics researches and the horizon of metabolomics researches could be recognized by reviewing the works that utilized and/or cited the KNApSAcK DB.

Methodologies for multivariate analysis to statistically process the massive amount of metabolome data were reviewed in [16] and to systematize blended herbal medicines in Kampo [15]. In the following section, we focus on the mining studies of blended herbal medicines for systematically understanding the composition of medicinal herbs to efficacies on humans, that is, principal component analysis (PCA) that makes it possible to systematize the ingredient in individual blending systems, partial least squares (PLS) that can relate the ingredients of medicinal herbs to the efficacies and N-PLS that can connect multi-factors to the efficacies. We initially explain individual techniques in Section 3 and then discuss their application in data-mining of blended types of herbal medicines in Section 4.

MATHEMATICAL METHODS OF DATA MINING

Principal Component Analysis (PCA)

PCA is a linear transformation of a large number of interrelated variables into a new set of variables, called as the principal components (PCs), which are uncorrelated and ordered so that the first few retain most of the variation present in all the original variables [132].

Consider a data matrix $A = (a_1 a_2$ 舰 $a_p)$ with n observations and let V ($p \times p$) be the variance-covariance matrix of A. The principal components of A, $Z = (z_1 z_2$ 舰 $z_p)$, are calculated as

$$z_j = Ac_j \ (j = 1, 2, \ldots, p) \tag{1}$$

where c_j is the j-th eigenvector of V which correspond to the j-th eigenvalue of V ($_j$). The properties of PCs are: (1) Var(zj) = $_j$; (2) Cov(z_j, z_J) = 0, $j \neq J$; (3) Var(z_1) $\geq$ Var(z_2) $\geq$ 舰 Var(z_p). The cumulative proportion of variance of the original variables explained by the first J principal components can be obtained as

$$\Pr(z_j) = \frac{\sum_{j=1}^{J} \lambda_j}{\sum_{j=1}^{p} \lambda_j} \quad (2)$$

Partial Least Squares

PLSR is a regression method, which assumes underlying factors among the predictors account for most of the response variation [133 and 134]. These underlying factors of X-variate

$$T = XW \quad (3)$$

are obtained by maximizing their covariance with the corresponding underlying factors of Y-variate where X is an n × m matrix of predictors, Y is an n × p matrix of responses, T is an n × c matrix of X-score factors, and W is m × c matrix of weight. Note that n is the number of observations, m is the number of predictors, p is the number of responses, and c is the number of components.

The X-score factors, i.e. matrix T, have the following properties [133].

a. When multiplied by loadings P, they are good summaries of X, i.e. the X-residuals E are small

$$X = TP^t + E \quad (4)$$

- The X-score factors are good predictors of **Y**, i.e.

$$Y = TQ^t + F \quad (5)$$

The Y-residuals F express the deviations between the observed and modeled responses.

Based on Eq. (3), Eq. (5) can be rewritten as a multiple regression model

$$Y = XWQ^t + F = XB + F \tag{6}$$

Thus, PLSR coefficients B can be written as

$$B = WQt \tag{7}$$

whereas prediction of the responses can be obtained from

$$\hat{Y} = XWQ^t \tag{8}$$

Although PLSR is not specifically designed to discriminate among groups, Barker and Rayens [135] have demonstrated that PLSR can be used for such purposes by connecting PLSR and Linear Discriminant Analysis (LDA); this combined method is called as Partial Least Square Discriminant Analysis (PLS-DA). In PLS-DA, group membership is transformed into a dummy matrix, and this dummy matrix provides the response variables for PLSR.

Multiway Model

An extension of PLSR to deal with multidimensional data known as Multiway Partial Least Squares has been developed by Bro [136] and is called as N-PLS. In this model, the same principle of PLSR for two dimensional data is utilized, that is, both predictor and response blocks are decomposed successively into multi-linear model such that the pairwise scores have maximal covariance. The score of the predictor is then regressed to the response variable. Fig. 2 illustrates the decomposition of N-PLS model. Moreover, N-PLS model can also be used for discrimination purpose, which is called as N-PLS-DA, that is the multiway version of PLS-DA, by utilizing the dummy matrix of group membership as the response variable.

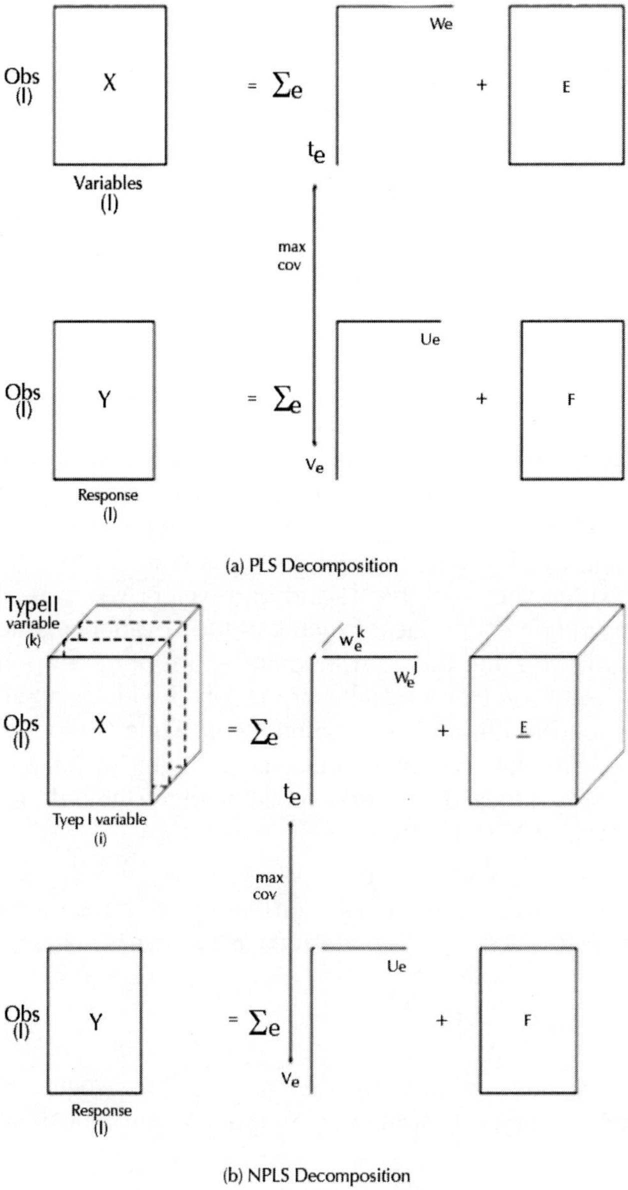

Figure 2: Schematic diagram of the decomposition of both predictor and response blocks for: (a) PLS and (b) N-PLS model.

Consider the three-dimensional array $\underline{X}$ indexed by observation (i = 1, 2, ..., I), type I variable (j = 1, 2, ..., J) and type II variable (k = 1, 2, ..., K). The decomposition of both the predictor and the response block based on N-PLS model are as follows

$$X_{ijk} = \sum_{c=1}^{C} T_{ic} W_{jc}^{J} W_{kc}^{K} + E_{ijk} \tag{9}$$

$$Y_{il} = \sum_{c=1}^{C} V_{ic} V_{lc} + F_{il} \tag{10}$$

The array $\underline{X}$ is decomposed into a tri-linear model consisting of one score vector for observation called tc (I× 1), and two weight vectors, one for type I variable called (J × 1) and one for type II variable called (K × 1). Similarly, a bi-linear model is used in decomposing the matrix Y into one score vector v_c (I × 1) and one weight vector u_c (L × 1). The decomposition is conducted such that the covariance among the score of predictor t and the corresponding score of the response v is maximized. All scores and weights are indexed with c showing that they correspond to cth multiway component, while C represents the total number of multiway components used in N-PLS model. Moreover, E and F are the residuals of the decomposition of the three-dimensional array $\underline{X}$ and matrix Y, respectively.

Furthermore, let X_k (I × J) be the kth slice of $\underline{X}$ (I × J × K) for the corresponding kth of type II variable, then matricizing three-dimensional array X into matrix X (I × JK) is performed as follows [137]

$$X = [X_1 | X_2 | \ldots |XK \tag{11}$$

Fig. 3 depicts this unfolding process of array $\underline{X}$ into matrix X. Using this notation, the score t_c of the cth component can be calculated as [138]

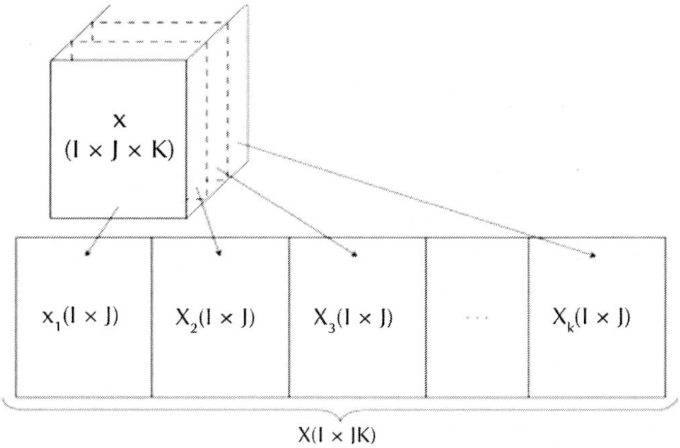

Figure 3: Illustration of matricizing three-dimensional array X (I × J × K) into matrix X̲ (I × J × K).

$$t_c = X\left(W_c^K \otimes W_c^J\right)$$

or

$$t_{ic} = \sum_{j=1}^{J}\sum_{k=1}^{K} x_{ijk} w_{jc}^J w_{kc}^K \tag{12}$$

From Eq. (12), the weight corresponding to cth component, w_c (JK × 1), can be defined as

$$W_c = \left(W_c^K \otimes W_c^J\right) \tag{13}$$

Smilde [140] also described that, due to the deflation in X during the decomposition, the weight matrix W(JK × C) can be applied directly to the original unfolded matrix X is defined as

$$W = \left[w_2 \middle| (I_{JK} - w_1 w_1^t) w_2 \middle| \ldots \middle| (I_{JK} - w_1 w_1^t)(I_{JK} - w_2 w_2^t) \ldots (I_{JK} - w_{Q-1} w_{Q-1}^t) w_Q \right] \tag{14}$$

Hence, the scores in **T** (I × C) expressed directly in terms of the X-columns is

$$T = XW \tag{15}$$

After the decomposition procedure, the next step is to regress Y on the component scores T

$$\hat{Y} = TB \tag{16}$$

with

$$B = (T^tT)^{-1}T^tY \tag{17}$$

From Eq. (15) and (16) we have

$$\hat{Y} = XWB \tag{18}$$

Therefore, the regression coefficients B_{NPLS} (JK × L) needed to predict Y from X are obtained as

$$B_{NPLS} = WB \tag{19}$$

ILLUSTRATION OF DATA MINING TECHNIQUES

Indonesia, the mega-biodiversity center like Brazil, has at least 9,600 species of plants with pharmacological activity [110] and has developed blended herbal medicines called Jamu taking modern and traditional knowledge of herbs into consideration. To prepare Jamu, several plants are selected and mixed such that the concoction has the desired efficacy. Traditionally, plants are chosen based on prior experience which is passed down from generation to generation. In curing a particular disease, each ethnic group in Indonesia may have its own formulas, whose specific nature depends strongly on the local plant resources in the region where a given population lives and the efficacies of Jamu medicines have been empirically demonstrated [139, 140, 141 and 142]. Data mining techniques with the blended herbal medicine databases such as KAMPO and JAMU (Fig. 1) makes it possible to comprehensively and mathematically understand those blended herbal systems. Fig. 4 illustrates a network connecting efficacy, herbal medicine, plant, and pharmacological activity of plant. The network showing that crude medicines M_1, which is useful for efficacy E_1, use three plants in its ingredients: plant P_1, P_3, and P_4. Plant P_1 has two pharmacological activities: A_2 and A_4. Plant P_2 also has two pharmacological activities: A_1 and A_2, while plant P_3 has three activities: A_3, A_4, and AK. The other connections can be described similarly.

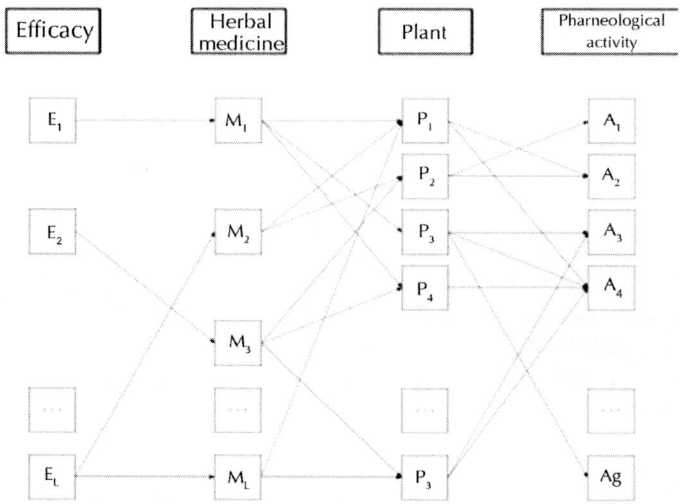

Figure 4: A typical network illustrating connections between efficacy, herbal medicine, plant, and pharmacological activity of plant.

From the concept of integrated platform of knowledge of medicinal plants and plant and human-omics depicted in Fig. 1, the efficacy layer in Fig. 4 represents the physiological activity layer in human-omics attribute, the herbal medicine and plant layer represent the prescription and medicinal herb layer, respectively, in knowledge of medicinal plants attribute, while the pharmacological activity layer represents the metabolomics layer in plant-omics attribute. On the following section we will illustrate the data mining techniques on herbal medicine database analyzing relationship among entities for two, and more than two attributes.

Two Attributes

As an illustration for data mining of herbal medicine database which rely on relationship between two attributes, the relationship between the efficacy of Jamu and medicinal plants used in Jamu is explored using PCA [143, 144 and 145]. The efficacies of 3,138 Jamu are classified into one of nine categories, namely: (1) disorders of appetite (DOA), (2) disorders of mood and behavior (DMB), (3) female reproductive organ problems (FML), (4) gastrointestinal disorders (GST), (5) musculoskeletal and connective tissue disorders (MSC), (6) pain/inflammation (PIN), (7) respiratory disease (RSP), (8) urinary related problems (URI), and (9) wounds and skin infections (WND). In total, those 3,138 Jamu use 465 plants in their ingredients. The distribution of Jamu and plant utilized in Jamu for each efficacy is shown inTable 2.

Table 2: Distribution of Jamu and plant utilized in Jamu for each efficacy

Efficacy	Number of Jamu	Number of plants utilized in Jamu formulas
Urinary-related problems (URI)	72	80
Disorders of appetite (DOA)	249	148
Disorders of mood and behavior (DMB)	22	47
Gastrointestinal disorders (GST)	980	290
Female reproductive organ problems (FML)	398	182
Musculoskeletal and connective tissue disorders (MSC)	840	270
Pain and inflammation (PIN)	311	183
Respiratory diseases (RSP)	107	105
Wounds and skin infection (WND)	159	120

Note that, one plant may be used in many Jamu with varying efficacies. Hence, it is interesting to find out the most significant effects of specific plants by analyzing their usage in Jamu, and considering that the more useful a given plant in having certain effect, the more frequently the plant will be used in Jamu when that effect is desired. Biplot, a multivariate exploration tool, is suitable for this purpose because it provides simultaneous plot of principal component scores and loadings, as representation of observations and variables, respectively [145]. Considering plants as observations and efficacy groups as variables, the relationship between them can be explored using a biplot.

Following the explanation of PCA in previous section, the data matrix A as an input for PCA is generated by putting plant as observation and efficacy as variables. So, A consists of 465 rows and 9 columns. Each cella_{ij} shows the number of Jamu that use plant i and useful for efficacy j.

Biplot configuration using the first two components is shown in Fig. 5. In the figure, plants are represented as red points while Jamu efficacies as blue lines, i.e. vectors based on loadings. The length of a given efficacy line showing the variability of plant usage for the corresponding efficacy, that is, the longer the efficacy line the larger the variability of plant usage for that efficacy. From Fig. 6, it is obvious that efficacy MSC has the largest variability of plant usage, followed by efficacy GST and FML. On the other hand, efficacy DMB has the smallest variability of plant usage, followed by efficacy URI and RSP. This finding can be addressed due to two factors, that is, the number of Jamu as well as the number of plant utilized in the corresponding efficacy (see Table 2). Efficacies with large variability of plants usage (MSC, GST, and FML) have large values for both factors; in contrast, efficacies with small variability of plants usage (efficacy DMB, URI, and RSP) have small values for both factors.

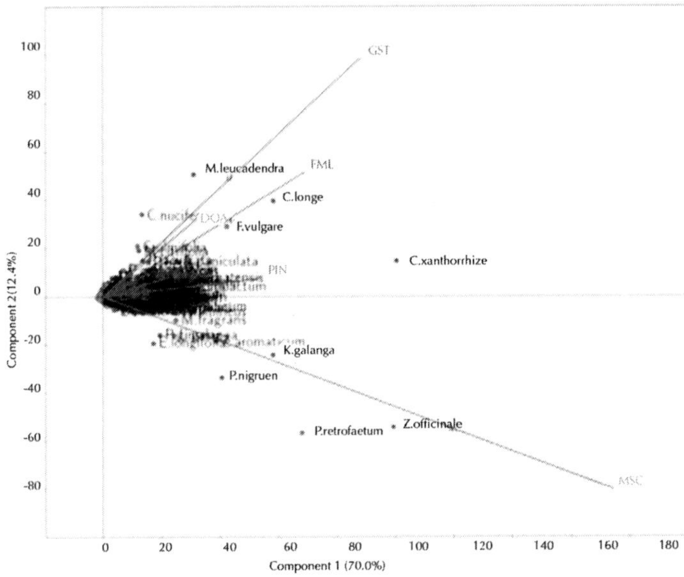

Figure 5: Biplot configuration based on PCA analysis of Jamu data. Plants and Jamu efficacies are represented as red points and blue lines, respectively.

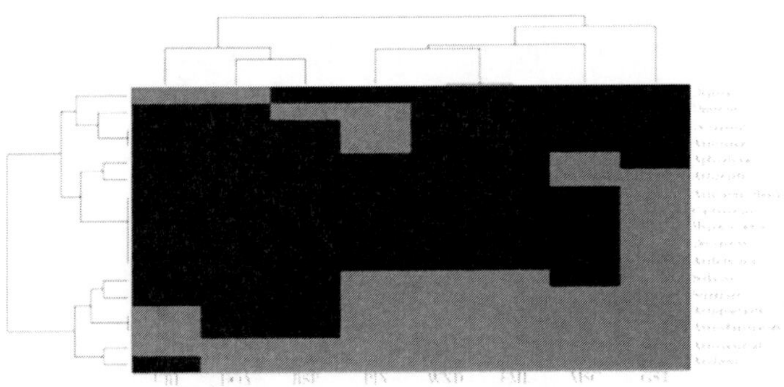

Figure 6: Clustergram of pharmacological activity against Jamu efficacy. The red and black cells indicate that the pharmacological activity is significant or non-significant, respectively, to the corresponding efficacy.

In the configurations, many plants are clustered in the center. Note that, the projection value of plants' point on a given efficacy line is the prediction of the frequency of plants usage on that efficacy. So, these clustered plants are basically plants whose frequencies of usage in Jamu are very low. In contrast to the clustered plants, some plants are spread out and located near the efficacy for which the plants are highly utilized. For example, Ginger (Zingiber officinale) is located near the efficacy MSC. Ginger is well known for its function of refreshing body, and for this reason many Jamu use Ginger for efficacy MSC which can easily be identified from biplot configuration. Another example is Turmeric (Curcuma longa) which located near the efficacy FML. Due to its analgesic and antimicrobial activity, this plant is well known and highly utilized in Indonesia as ingredient of Jamu formula for women during menstruation, which is a problem that classified into efficacy FML. Thus, the biplot configuration exhibits useful information in exploring the relationship between plants and the efficacy of Jamu.

Another illustration for relationship between two attributes on data mining of herbal medicine database is the modeling of Jamu ingredients (representation of knowledge of medicinal plants) to predict the efficacy (representation of human omics). This analysis is performed because of the fact that Jamu is prepared from a mixture of several plants. The plants are chosen so that the Jamu has the desired efficacy. As a result, the composition of the plants used in Jamu formula determines the efficacy. Thus, it is interesting to model the ingredients of Jamu, i.e. the constituent plants, and use this model to predict efficacy. PLS-DA, a statistical model for classification and discrimination based on Partial Least Square Regression (PLSR), is suitable for this analysis because a large number of plants are used in Jamu, whereas Jamu efficacies can be grouped into a few categories or classes. In this method, the plants used in each Jamu medicine served as the predictors, whereas the efficacy of each Jamu provided the responses.

The data structure used for PLS-DA is as follows. The data matrix X in X-block contains plant usage status. The dimension of matrix X is $(I \times J)$, where I is the number of Jamu (in this case, 3,138), and J is the number of plants (in this case, 465). Because of the availability of information about Jamu products, which generally do not state in detail the mixing ratio of the plants used, the predictors X is constructed only in binary data. Each cell x_{ij} $(i = 1, 2, \ldots, I; j = 1, 2, \ldots, J)$ is set to 1 if Jamu i uses plant j, and is set to 0 otherwise. In the present study, nine indicator variables, which correspond to the 9 efficacies listed in Table 2 perform as the Y-block in PLS-DA modeling. Thus, the dimension of data matrix Y is (I × 9). Each cell y_{il} (l= 1, 2, ..., 9) is set to 1 if Jamu i is classified into efficacy group 1, and is set to 0 otherwise. Note that $\sum_{l=1}^{9} y_{il} = 1$ because each Jamu is classified to one efficacy only.

Using the derived PLS-DA model, we can then use it to predict the efficacy of Jamu given information of the ingredients. In this analysis, among the 3,138 Jamu medicines, the efficacies of 2,248 Jamu medicines (71.6%) can be assigned to an individual efficacy reported. Hence, the efficacy in most Jamu medicines can be predicted on the basis of medicinal plants used. The percentages of correct prediction for each efficacy (see Table 3) vary from 22.7% for efficacy DMB to 89.8% for efficacy GST. The low percentage of correct prediction for efficacy DMB can be addressed due to the small number of Jamu for this efficacy, which is only 22 out of 3,138 Jamu (see Table 2).

Table 3: Confusion matrix of the prediction of Jamu efficacy using the PLS-DA model

Observed efficacy	Predicted efficacy										
	URI	DOA	DMB	GST	FML	MSC	PIN	RSP	WND	Total	% Correct
URI	39	0	0	21	2	10	0	0	0	72	54.2
DOA	0	164	0	29	36	18	0	0	2	249	65.9
DMB	0	1	5	10	0	3	1	2	0	22	22.7
GST	3	17	0	880	12	46	9	6	7	980	89.8
FML	0	13	0	61	266	50	5	1	2	398	66.8
MSC	6	6	1	127	41	638	16	0	5	840	76
PIN	1	0	0	90	4	77	133	4	2	311	42.8
RSP	3	0	0	21	4	23	3	52	1	107	48.6
WND	2	3	0	57	11	11	4	0	71	159	44.7
Total	54	204	6	1296	376	876	171	65	90	3138	71.6

Furthermore, plants in the ingredients of Jamu are used as main ingredients, which contribute primarily to the medicines' efficacies; other plants are used as supporting ingredients [146 and 147]. Investigating which plants are main ingredients and which are supporting is important in order to comprehensively understand the mechanisms by which specific plants achieve desired efficacies. The regression coefficients of previous PLS-DA model, which relates plants usage in Jamu as predictors and Jamu efficacy as response, can be helpful in this attempt because they summarize the effect of plant on efficacy. Plants that act as main ingredients will have significant effect on the model developed. Furthermore, due to the absence of parametric testing for the PLS-DA coefficients, the evaluation for significance is performed using permutation testing, in which the distribution of coefficients under the null hypothesis is generated via resampling of the existing data [149].

The resampling is performed by permuting the order of the responses (in this case, Jamu efficacies) while maintaining the order of the predictors (in this case, plant utilization as Jamu ingredients) so that the existing relationship between the predictors and the response is destroyed and a new data set is generated under the null hypothesis, i.e., plant utilization in Jamu does not affect Jamu efficacy. If we perform such resampling many times and apply the PLS-DA model on the new data generated from the resampling, the accumulation of the PLS-DA coefficients obtained from this process generates a distribution, against which a p-value can be calculated and subsequently evaluated for significance [150].

The results of the significance testing of all plants used in each 9 efficacies are shown in Table 4. Note that one plant may be used for more than one efficacy. From the testing, we observed 234 plants (50.3% among all 465 plants) showing no significant status for all 9 efficacies; whereas the other 231 plants have significant status which comprise of 189 plants (40.6%) are significant only for 1 efficacy, 38 plants (8.2%) are significant for 2 efficacies, and the other 4 plants (0.9%) are significant for 3 efficacies. Besides testing the plants usage statistically, furthermore, we also checked from scientific papers the usage of significant plants in their corresponding efficacy. Many of the results we obtained by our analysis are supported by scientific papers.

Table 4: Number of significant plants for each efficacy

Efficacy	Total	Support from scientific paper	
URI	20	15	−75.00%
DOA	21	20	−95.20%
DMB	12	6	−50.00%
GST	26	23	−88.50%
FML	40	30	−75.00%

MSC	40	39	−97.50%
PIN	39	37	−94.90%
RSP	36	33	−91.70%
WND	43	38	−88.40%

Note that in predicting Jamu efficacy based on the information of its ingredients we can also use other methods such as discrimination analysis, nominal logistic regression, and support vector machine. However, in the present study we focus on PLS-DA in classifying Jamu efficacy by taking into consideration that we also intend to evaluate the significance of plant usage in Jamu to achieve specific efficacy as well as extending the analysis into three-way model by adding the plant pharmacological activity into predictors' block.

More than Two Attributes

During the modeling process of PLS-DA in the previous section, the ingredients of Jamu provide the predictor while the Jamu efficacy serves as the response. In order to identify the function of the plants in Jamu to achieve specific efficacy, the reported pharmacological activities of the plants are added to the predictors block. Thus, the predictors block can be represented as a three-dimensional array X (I × J × K) indexed by Jamu medicine (i), plant (j), and pharmacological activity (k) as depicted in Fig. 2 with Jamu medicine, plant, and pharmacological activity serve as observation, type I and type II variables, respectively. Furthermore, the response block is represented as matrix Y (I × 9). This analysis then connects three attributes: (1) knowledge of medicinal plants (represented by Jamu and plants corresponding to JAMU DB in Fig 1); (2) plant omics (represented by pharmacological activity corresponding to Biological activity (Nat) in Fig 1); and (3) human omics (represented by efficacy).

The detail about the elements of array $\underline{X}$ and matrix Y is as the following. Let x_{ijk} (k = 1, 2, …;, K; K = 46 where K is the number of reported pharmacological activity; see previous section on definition of i, j, I, andJ) denotes the usage status of plant j with pharmacological activity k in Jamu i, where x_{ijk} = 1 if the plant j with pharmacological activity k is used in Jamu i, and x_{ijk} = 0 otherwise. On the other hand, let y_{il} represents the status of Jamu i on efficacy l, where $y_{il} = 1$ if Jamu i is classified into efficacy l, and $y_{il} = 0$ otherwise.

In order to identify the pharmacological activity that is significantly related with the efficacy, we adopt the guidelines from Hair et al. [150] that all weights w^K (in absolute values) of 0.3 or above are significant for sample sizes of 350 or greater. Figure 6 depicts the 2-dimensional dendrogram of Jamu efficacy and the pharmacological activity significantly related with the efficacy. The cluster of Jamu efficacy and the pharmacological activity was performed using Ward Linkage based on the Euclidean distance among the entities. The clustering of the pharmacological activity side clearly exhibits two groups. The first group consists of activities useful for one or two efficacies only. This group can be regarded as a group of specific activity because the effects of the activities are specific for certain efficacy. For example the diuretic activity is useful for efficacy URI and DOA. Diuretic is an agent that increases the secretion and elimination of urine from the body [151]. Obviously, this activity is beneficial for the efficacy URI. Diuretic also help the body eliminate waste and support the whole process of inner cleansing, which is an action that is useful for efficacy DOA especially related with a slimming purpose. The five activities (antihaemorrhoidal, carminative, hypoglycaemic, depurative, and anthelmintic) are specifically related with efficacy GST. Antihaemorrhoidal means an activity that treats haemorrhoids (piles), while the carminative is defined as an activity that eases discomfort caused by flatulence. Hypoglycaemic activity helps reduce the levels of sugar in the blood, whereas the depurative eliminates toxins and purifies the system especially the blood, and the anthelmintic helpful in expelling parasites from the gut. Thus, all of these activities are helpful for the problem related with the digestive system, i.e. the efficacy GST.

Furthermore, the second group of activity revealed by the dendrogram consists of activities useful for at least four efficacies. In contrast to the first group, this group can be regarded as the general activities because of the diverse efficacies related to this group. Among all activities clustered to this group, antimicrobial activity is significantly related with all 8 efficacies. We can interpret this result as follows. Due to the environmental conditions, hygiene, and its location as a tropical country which led to many microbes that are harmful to health, then it is reasonable that antimicrobial activity is important and should be available in many Jamu formulas in Indonesia. It should be noted that many popular medicinal plants in Indonesia such as Temulawak (Curcuma xanthorriza), Ginger (Zingiber officinale), Turmeric (Curcuma longa) or Kencur (Kaempferia galanga) have content of this activity [152].

Anti-inflammation, antispasmodic, analgesic, sedative, and stimulant are also clustered into this general activity group. Since many health problems or diseases are often accompanied with inflammation or spasm, then the plants with anti-inflammation and/or antispasmodic activity are chosen in many Jamu formulas. Those health problems/diseases often cause pain or other discomforts, thus plants with certain activities such as analgesic or sedative effects are chosen in many Jamu medicines. Finally, stimulant activity, which excites or quickens activity of the physiological processes, is important for the recovery reason after one experiencing those health problems or diseases.

From the previous explanation regarding the grouping of pharmacological activity, it can be concluded that in formulating Jamu the plants are selected so that, beside curing the targeted diseases or health problems as indicated by the specific activities, the plants also should overcome the other discomforts caused by the targeted diseases or health problems as indicated by the general activities. It is in accordance with the process of making the Jamu medicines that involving whole part of plant and not only the specific active components. Hence specific or general pharmacological activities of components are involved during the curing process of Jamu medicines towards targeted diseases or health problems.

CONCLUDING REMARKS

Biology, like most scientific disciplines, is in an era of accelerated information gathering and scientists increasingly depend on the availability of amounts of data such as nucleotide and protein sequences, protein and gene expression, dynamics of metabolites etc. The nature of current systematic understanding of big data biology towards health, nutrition, and other societal issues have recently become the focus of scholar in societal studies of science and information studies. The rise of community databases, i.e., KNApSAcK family DB introduced in the present review, has been strongly associated with the current emphasis on data-intensive science. The central question is whether scientists can deduce how systems and whole organisms work from this torrent of molecular data. To progress this situation, data-intensive approach is needed for understanding intra- and inter-relations in individual layers represented in Fig. 1. The former can be solved based on a type of multivariate analyses such as cluster analysis and principal component analysis. Though the latter is more complicated, several approaches including PLS and N-PLS make it possible to clarify and understand those relations. The big data biology has become an inevitable part of biology, and the laws of nature could be clarified based on global analysis of big data biology the era of which has appeared. For centuries biological research mainly depended on experiments and for a decade or two computational analysis has usually followed experimentation but future it might be the opposite i.e., computational analysis is done first to guide the experimental design facilitated by versatile and freely available omics data at various databases.

ACKNOWLEDGEMENTS

This work was partially supported by the National Bioscience Database Center in Japan, the Ministry of Education, Culture, Sports, Science and Technology of Japan (Grant-in-Aid for Scientific Research on Innovation Areas "Biosynthetic Machinery. Deciphering and Regulating the System for Creating Structural Diversity of Bioactive Metabolites (2007)") and CREST.

REFERENCES

1. MJ Raddick, AS Szalay, The universe online, Sci., 329 (2010), pp. 1028–1029
2. W Callebaut, Scientific perspectivism: A philosopher of science's response to the challenge of big data biology, Studies History Philosophy Biol. Biomed. Sci., 43 (2012), pp. 69–80
3. E Aronova, KS Baker, N Oreskes, Big science and big data in biology: From the international geophysical year through the international biological program to the long term ecological research (LTER) network, 1957-present, Historical Studies Natural Sci., 40 (2010), pp. 183–224

4. CH Liu, DY Wu, JD Pollock, Bioinformatic challenges of big data in non-coding RNA research, General Commentary, 3 (2012), pp. 1–3
5. AE Thessen, DJ Patterson, Data issues in the life sciences, Zookeys, 150 (2011), pp. 15–51
6. E Pennisi, How will big pictures emerge from a sea of biological data, Sci., 309 (2005), p. 94
7. S Ranganathan, C Schonbach, J Kelso, B Rost, S Nathan, TW Tan, Towards big data science in the decade ahead from ten years of InCoB and the 1st ISCB-Asia Joint Conference, BMC Bioinf., 12 (2011), pp. 51.1–51.4
8. E Birney, Lessons for big-data project, Nature, 489 (2012), p. 49
9. RF Service, Materials scientists look to a data-intensive future, Sci., 335 (2012), pp. 1434–1435
10. WK Michener, MB Jones, Ecoinformatics: supporting ecology as a data-intensive science, Trends Ecol. Evol., 27 (2012), pp. 85–93
11. WM Hochachka, D Fink, RA Hutchinson, D Sheldon, WK Wong, S Kelling, Data-intensive science applied to broad-scale citizen science, Trend Ecol. Evol., 27 (2011), pp. 130–137
12. EE Schadt, The changing privacy landscape in the era of big data, Mol. Sys. Biol., 8 (2012), pp. 1–3
13. Verpoorte R, Kim HK & Choi YH (2006) Plants as source of medicines, Medicinal and Aromatic Plants Chapter 19, Edited by Boger RJ, Craker LE & Lange D
14. FM Afendi, T Okada, M Yamazaki, A Hirai-Morita, Y Nakamura, K Nakamura, S Ikeda, H Takahashi, MAF Amin, LK Daruman, K Saito, S Kanaya, KNApSAcK family databases: Integrated metabolite-plant species databases for multifaceted plant research, Plant Cell Physiol., 53 (2011), pp. e1.1–e112
15. FM Afendi, T Katsuragi, A Kato, N Nishihara, K Nakamura, Y Nakamura, K Tanaka, A Hirai-Morita, MAU Amin, H Takahashi, S Kanaya, Systems biology approaches and metabolomics for understanding Japanese traditional Kampo medicine, Curr Pharm Personalized Med., 10 (2012), pp. 111–124
16. T Okada, FM Afendi, MAU Amin, H Takahashi, K Nakamura, S Kanaya, Metabolomics of medicinal plants: The importance of

multivariate analysis of Analytical chemistry data, Curr Computer-Aided Drug Design, 6 (2010), pp. 179–196

17. C Ash, P Kiberstis, E Marshall, J Travis, It takes more than an apple a day, Sci., 337 (2012), p. 1467

18. RJ Binoa, RD Hall, O Fiehn, J Kopka, K Saito, J Draper, BJ Nikolau, P Mendes, U Roessner-Tunali, M Beale, RN Trethewey, BM Lange, ES Wurtele, LW Sumner, Potential of metabolomics as a functional genomics tool, Trends Plant Sci., 9 (2004), pp. 418–425

19. M Macel, NM van Dam, JB Keurentjes, Metabolomics: the chemistry between ecology and genetics, Mol. Ecol. Resources, 10 (2010), pp. 583–593

20. K Saito, F Matsuda, Metabolomics for functional genomics, systems biology, and biotechnology, Annu. Rev. Plant. Biol., 61 (2010), pp. 463–489

21. R Verpoorte, YH Choi, HK Kim, Ethnopharmacology and systems biology: a perfect holistic match, J. Ethnopharmacol., 100 (2005), pp. 53–56

22. LF Shyur, NS Yang, Metabolomics for phytomedicine research and drug development, Curr. Opin. Chem. Biol., 12 (2008), pp. 66–71

23. Y Shinbo, Y Nakamura, M Altaf-Ul-Amin, H Asahi, K Kurokawa, M Arita, K Saito, D Ohta, D Shibata, S Kanaya, KNApSAcK: a comprehensive species-metabolite relationship database, K Saito, RA Dixon, L Willmitzer (Eds.), Biotechnology in agriculture and forestry 57. Plant metabolomics, Springer, Berlin (2006), pp. 165–181

24. K. Kikuchi, H Kakeya, A bridge between chemistry and biology, Nature Chem. Biol., 2 (2006), pp. 392–394

25. A Oikawa, Y Nakamura, T Ogura, A Kimura, H Suzuki, N Sakurai, Y Shinbo, D Shibata, S Kanaya, D Ohta, Clarification of pathway-specific inhibition by Fourier transform ion cyclotron resonance/mass spectrometry-based metabolic phenotyping studies, Plant Physiol., 142 (2006), pp. 398–413

26. Y Shinbo, S Sakaguchi, Y Nakamura, M Altaf-Ul-Amin, K Kurokawa, K Funatsu, S Kanaya, pecies-metabolite Database

(KNApSAcK): Elucidating Diversity of Flavonoids, SComput. Aided Chem., 7 (2006), pp. 94–101

27. T Tomiki, T Saito, M Ueki, M Konno, T Asaoka, R Suzuki, M Uramoto, H Kakeya, H Osada RIKEN Natural Products Encyclopedia (RIKEN NPEdia), a Chemical Database of RIKEN Natural Products Depository (RIKEN NPDepo), Comput. Aided Chem., 7 (2006), pp. 157–162
28. A. Gaida, SJ Neumann, MetHouse: Raw and Preprocessed Mass Spectrometry Data J Integrative Bioinformatics, 4 (2007), pp. 1–8
29. J Hummel, J Selbig, D Walther, J Kopka, The Golm metabolome database: a database for GC-MS based metabolite profiling, Topics Curr Genet., 18 (2007), pp. 75–95
30. D Ohta, D Shibata, S Kanaya, Metabolic profiling using Fourier-transform ion-cyclotron-resonance mass spectrometry, Anal. Bioanal. Chem., 389 (2007), pp. 1469–1475
31. S Moco, J Vervoort, CHR de Vos, RJ Bino, Metabolomics technologies and metabolite identification, Trends Anal. Chem., 26 (2007), pp. 855–866
32. K Saito, MY Hirai, K Yonekura-Sakakibara, Decoding genes with coexpression networks and metabolomics – 'majority report by precogs', Trends Plant Sci., 13 (2008), pp. 36–43
33. K Yonekura-Sakakibara, T Tohge, R Niida, K Saito, Identification of a Flavonol 7-O-Rhamnosyltransferase Gene Determining Flavonoid Pattern in Arabidopsis by Transcriptome Coexpression Analysis and Reverse Genetics, J. Biol. Chem., 282 (2007), pp. 14932–14941
34. EJ Want, A Nordstrom, H Morita, G Siuzdak, From Exogenous to Endogenous: The Inevitable Imprint of Mass Spectrometry in Metabolomics, J. Proteome Res., 6 (2007), pp. 459–468
35. Y Nakamura, A Kimura, H Saga, A Oikawa, Y Shinbo, K Kai, N Sakurai, H Suzuki, M Kitayama, D Shibata, S Kanaya, D Ohta, Differential metabolomics unraveling light/dark regulation of metabolic activities in Arabidopsis cell culture, Planta, 227 (2007), pp. 57–66
36. K Akiyama, E Chikayama, H Yuasa, Y Shimada, T Tohge, K Shinozaki, MY. Hirai, T Sakurai, J. Kikuchi, K Saito, PRIMe: a Web

site that assembles tools for metabolomics and transcriptomics, In Silico Biol., 8 (2008), pp. 339–345

37. C Bottcher, E von Roepenack-Lahaye, J Schmidt, C Schmotz, S Neumann, D Scheel, S Clemens, Metabolome Analysis of Biosynthetic Mutants Reveals a Diversity of Metabolic Changes and Allows Identification of a Large Number of New Compounds in Arabidopsis, Plant Physiol., 147 (2008), pp. 2107–2120
38. WB Dunn, Current trends and future requirements for the mass spectrometric investigation of microbial, mammalian and plant metabolomes, Phys. Biol., 5 (2008), pp. 1–24
39. A Fardet, R Llorach, A Orsoni, JF Martin, E Pujos-Guyot, C Lapierre, A Scalbert, Metabolomics provide new insights on the metabolism of dietary phytochemicals in rats, J. Nutr., 138 (2008), pp. 1282–1287
40. A Fait, K Hanhineva, R Beleggia, N Dai, I Rogachev, VJ Nikiforova, AR Fernie, A Aharoni, Reconfiguration of the achene and receptacle metabolic networks during strawberry fruit development, Plant Physiol., 148 (2008), pp. 730–750
41. P Giavalisco, J Hummel, J Lisec, AC Inostroza, G Catchpole, L Willmitzer, High-resolution direct infusion-based mass spectrometry in combination with whole 13C metabolome isotope labeling allows unambiguous assignment of chemical sum formulas, Anal. Chem., 80 (2008), pp. 9417–9425
42. JM Hagel, P Facchini, Plant metabolomics: analytical platforms and integration with functional genomics, Phytochem. Rev., 7 (2008), pp. 479–497
43. K Hanhineva, I Rogachev, H Kokko, S Mintz-Oron, I Venger, S Karenlampi, A Aharoni, Non-targeted analysis of spatial metabolite composition in strawberry (Fragaria x ananassa) flowers, Phytochemistry, 69 (2008), pp. 2469–2481
44. K Hanhineva, Metabolic Engineering of Phenolic Biosynthesis Pathway and Metabolite Profiling of Strawberry (Fragaria x ananassa), Doctoral dissertationUniv. of Kuopio (2008)
45. Y Iijima, Y Nakamura, Y Ogata, K Tanaka, N Sakurai, K Suda, T Suzuki, H Suzuki, K Okazaki, M Kitayama, S Kanaya, K Aoki, D Shibata, Metabolite annotations based on the integration of mass spectral information, Plant J, 54 (2008), pp. 949–962

46. S Malitsky, E Blum, H Less, I Venger, M Elbaz, S Morin, Y Eshed, A Aharoni, The transcript and metabolite networks affected by the two clades of Arabidopsis glucosinolate biosynthesis regulator, Plant Physiol, 148 (2008), pp. 2021–2049

47. F Matsuda, K Yonekura-Sakakibara, R Niida, T Kuromori, K Shinozaki, K Saito, MS/MS spectral tag-based annotation of non-targeted profile of plant secondary metabolites, Plant J, 57 (2008), pp. 555–577

48. S Mintz-Oron, T Mandel, I Rogachev, L Feldberg, O Lotan, M Yativ, Z Wang, R Jetter, I Venger, A Adato, A Aharoni, Gene expression and metabolism in tomato fruit surface tissues, Plant Physiol., 147 (2008), pp. 823–851

49. A Oikawa, F Matsuda, M Kusano, Y Okazaki, K Saito, Rice metabolomics, Rice, 1 (2008), pp. 63–71

50. DP Overy, DP Enot, K Tailliart, H Jenkins, D Parker, M Beckmann &, J Draper, Explanatory signal interpretation and metabolite identification strategies for nominal mass FIE-MS metabolite fingerprints. Nature Protochols, 3 (2008), pp. 471–485

51. H Takahashi, K Kai, Y Shinbo, K Tanaka, D Ohta, T Oshima, M Altaf-Ul-Amin, K Kurokawa, N Ogasawara, S Kanaya, Metabolomics approach for determining growth-specific metabolites based on Fourier transform ion cyclotron resonance mass spectrometry, Anal. Bioanal. Chem., 391 (2008), pp. 2769–2782

52. E Werner, JF Heilier, C Ducruix, E Ezan, C Junot, JC Tabet, Mass spectrometry for the identification of the discriminating signals from metabolomics: current status and future trends. J. Chromatogr. B., 871 (2008), pp. 143–163

53. T Ara, N Sakurai, Y Tange, Y Morishita, H Suzuki, K Aoki, K Saito, D Shibata, Improvement of the quantitative differential metabolome pipeline for gas chromatography-mass spectrometry data by automated reliable peak selection, Plant Biotechnol., 26 (2009), pp. 445–449

54. M Arita, K Suwa, Search extension transforms Wiki into a relational system: a case for flavonoid metabolite database, BMC BioData Mining, 1 (2009), pp. 7.1–7.8

55. K Bando, T Kunimatsu, J Sakai, J Kimura, H Funabashi, T Seki, T Bamba, E Fukusaki, GC-MS-based metabolomics reveals

mechanism of action for hydrazine induced hepatotoxicity in rats, Appl. Toxicol., 31 (2011), pp. 524–535

56. MP Davey, FI Woodward, WP Quick, Intraspecfic variation in cold-temperature metabolic phenotypes of Arabidopsis lyrata ssp petraea, Metabolomics, 5 (2009), pp. 138–149
57. J Draper, DP Enot, D Parker, M Beckmann, S Snowdon, W Lin, H Zubair, Metabolite signal identification in accurate mass metabolomics data with MZedDB, an interactive m/z annotation tool utilising predicted ionisation behaviour 'rules', BMC Bioinformatics, 10 (2009), pp. 277.1–27716
58. A Fukushima, M Kusano, H Redestig, M Arita, K Saito, Integrated omics approaches in plant systems biology, Curr. Opin. Chem. Biol., 13 (2009), pp. 532–538
59. J Han, R Datla, S Chan, CH Borchers, Mass spectrometry-based technologies for high-throughput metabolomics, Bioanalysis, 1 (2009), pp. 1665–1684
60. N Hounsome, B Hounsome, D Tomos, G Edward-Jones, Changes in antioxidant compounds in white cabbage during winter storage, Postharvest Biol. Tech., 52 (2009), pp. 173–179
61. T Kind, M Scholz, O Fiehn, How large is the metabolome? A critical analysis of data exchange practices in chemistry, PLoS One, 4 (2009), pp. e5440.1–e544010
62. K Kai, H Hashidzume, K Yoshimura, H Suzuki, N Sakurai, D Shibata, D Ohta, P450-dependent fatty acid hydroxylation reactions in Arabidopsis, Plant Biotechnol., 26 (2009), pp. 175–182
63. C Manach, J Hubert, R Llorach, A Scalbert, The complex links between dietary phytochemicals and human health deciphered by metabolomics, Mol. Nutr. Food Res., 53 (2009), pp. 1303–1315
64. F Matsuda, Y Shinbo, A Oikawa, MY Hirai, O Fiehn, S Kanaya, K Saito, Assessment of metabolome annotation quality: a method for evaluating the false discovery rate of elemental composition searches, PLoS One, 4 (2009), pp. e7490.1–e749010
65. F Matsuda, H Redestig, Y Sawada, Y Shinbo, MY Hirai, S Kanaya, K Saito, Visualization of metabolite identifier information, Plant Biotechnol., 26 (2009), pp. 479–483

66. R Oishi, K Tanaka, T Hashimoto, Y Shinbo, K Jumtee, T Bamba, E Fukusaki, H Suzuki, D Shibata, H Takahashi, H Asahi, K Kurokawa, Y Nakamura, A Hirai, K Nakamura, M Altaf-Ul-Amin, S Kanaya, An approach to peak detection in GC-MS chromatograms and application of KNApSAcK database in prediction of candidate metabolites, Plant Biotechnol., 26 (2009), pp. 167–174

67. T Okada, Y Nakamura, S Kanaya, A Takano, KJ Malla, T Nakane, M Kitayama, S Sekita, Metabolome analysis of Ephedra plants with different contents of ephedrine alkaloids by using UPLC-Q-TOF-MS, Planta Med., 75 (2009), pp. 1352–1356

68. Y Sawada, K Akiyama, A Sakata, A Kuwahara, H Otsuki, T Sakurai, K Saito, MY Hirai, Widely targeted metabolomics based on large-scale MS/MS data for elucidating metabolite accumulation patterns in plants. Plant Cell Physiol., 50 (2009), pp. 37–47

69. R Shroff, L Rulisek, J Doubsky, A Svatos, Acid-base-driven matrix-assisted mass spectrometry for targeted metabolomics, Proc. Natl. Acad. Sci. USA, 106 (2009), pp. 10092–10096

70. R Stracke, RC De Vos, L Bartelniewoehner, H Ishihara, M Sagasser, S Martens, B Weisshaar, Metabolomic and genetic analyses of flavonol synthesis in Arabidopsis thaliana support the in vivo involvement of leucoanthocyanidin dioxygenase, Planta, 229 (2009), pp. 427–445

71. K Takemoto, M Arita, Heterogeneous distribution of metabolites across plant species, Physica A, 388 (2009), pp. 2771–2780

72. K. Tanaka, K. Nakamura, T. Saito, H Osada, A Hirai, H Takahashi, S Kanaya, M Altaf-Ul-Amin, Metabolic pathway prediction based on inclusive relation between cyclic substructures, Plant Biotechnol., 26 (2009), pp. 459–468

73. K Tanaka, A Ina, Y Ohta, Comparative study of chemical constituents of the traditional medicine hochuekkito by LC-MS with multivariate statistical analysis, J. Trad. Med., 26 (2009), pp. 179–186

74. S Tianniam, T Bamba, E Fukusaki, Non-targeted metabolite fingerprinting of Oriental folk medicine Angelica acutiloba roots by ultra performance liquid chromatography time-of-flight mass spectrometry, J. Sep. Sci., 32 (2009), pp. 2233–2244

75. T Tohge, A Fernie, Web-based resources for mass-spectrometry-based metabolomics: a user's guide. Phytochemistry, 70 (2009), pp. 450–456

76. DS Wishart, Computational strategies for metabolite identification in metabolomics, Bioanalysis, 1 (2009), pp. 1579–1596

77. Z Xie, X Ma, DR Gang, Modules of co-regulated metabolites in turmeric (Curcuma longa) rhizome suggest the existence of biosynthetic modules in plant specialized metabolism, J. Exp. Botany, 60 (2009), pp. 87–97

78. K Yonekura-Sakakibara, K Saito, Functional genomics for plant natural product biosynthesis, Nat. Prod. Rep., 26 (2009), pp. 1466–1487

79. KA Aliferis, S Jabaji, Metabolite composition and bioactivity of Rhizoctonia solani sclerotial exudates, J. Agric. Food Chem., 58 (2010), pp. 7604–7615

80. V Bollina, GK Kumaraswamy, AC Kushalappa, TM Choo, Y Dion, S Rioux, D Faubert, H Hamzehzarghani, Mass spectrometry based metabolomics application to identify quantitative resistance related metabolites in barley against Fusarium head blight, Mol. Plant Pathol., 11 (2010), pp. 769–782

81. A Bar-Akiva, R Ovadia, I Rogachev, C Bar-Or, E Bar, Z Freiman, A Nissim-Levi, N Gollop, E Lewinsohn, A Aharoni, D Weiss, H Koltai, M Oren-Shamir, Metabolic networking in Brunfelsia calycina petals after flower opening, J. Exp. Botany, 61 (2010), pp. 1393–1403

82. M Hattori, N Tanaka, M Kanehisa, S Goto, SIMCOMP/SUBCOMP: chemical structure search servers for network analyses, Nucl. Acids Res., 38 (2010), pp. W652–W656

83. H Horai, M Arita, S Kanaya, Y Nihei, T Ikeda, K Suwa, Y Ojima, K Tanaka, S Tanaka, K Aoshima, Y Oda, Y Kakazu, M Kusano, T Tohge, F Matsuda, Y Sawada, MY Yokota Hirai, H Nakanishi, K Ikeda, N Akimoto, T Maoka, H Takahashi, T Ara, N Sakurai, H Suzuki, D Shibata, S Neumann, T Iida, K Tanaka, K Funatsu, F Matsuura, T Soga, R Taguchi, K Saito, T Nishioka, MassBank: a public repository for sharing mass spectral data for life sciences. Mass Spectrometry, 45 (2010), pp. 702–714

84. T Kind, O Fiehn, Advances in structure elucidation of small molecules using mass spectrometry, Bioanal. Rev., 2 (2010), pp. 23–60

85. M Macel, NM. Van Dam, JJB Keurentjes, Metabolomics: the chemistry between ecology and genetics, Mol. Ecol. Resources, 10 (2010), pp. 583–593

86. F Matsuda, MY Hirai, E Sasaki, K Akiyama, K Yonekura-Sakakibara, NJ Provart, T Sakurai, Y Shimada, K Saito, AtMetExpress Development: a phytochemical atlas of Arabidopsis development, Plant Physiol., 152 (2010), pp. 566–578

87. S Neumann, S Bocker, Computational mass spectrometry for metabolomics – a review, Anal. Biol. Chem., 398 (2010), pp. 2779–2788

88. Neveu V, Perez-Jimenez J, Vos F, Crespy V, du Chaffaut L, Mennen L, Knox C, E isner R, Cruz J, Wishart D & Scalbert A (2010), Database, (2010) Phenol-Explorer: an online comprehensive database on polyphenol contents in foods, doi:10.1093/database/bap024

89. D Ohta, S Kanaya, H Suzuki, Application of Fourier-transform ion cyclotron resonance mass spectrometry to metabolic profiling and metabolite identification, Curr. Opinion in Biotechnol., 21 (2010), pp. 35–44

90. L Penn, H Boeing, CJ Boushey, LO Dragsted, J Kaput, A Scalbert, AA Welch, JC Mathers, Assessment of dietary intake: NuGO symposium report, Genes Nutr., 5 (2010), pp. 205–213

91. H Redestig, M Kusano, A Fukushima, F Matsuda, K Saito, M Arita, Consolidating metabolite identifiers to enable contextual and multi-platform metabolomics data analysis, BMC Bioinformatics, 11 (2010), pp. 214.1–21411

92. K Saito, F Matsuda, Metabolomics for functional genomics, systems biology, and biotechnology, Annu. Rev. Plant Biol., 61 (2010), pp. 463–489

93. D Singla, A Sharma, J Kaur, B Panwar, GP Raghava, BIAdb: a curated database of benzylisoquinoline alkaloids, BMC Pharmacol., 10 (2010), pp. 4.1–4.8

94. K Tanaka, A Ina, K Hayashi, K Komatsu, Comparison of Glycyrrhizae Radix from various sources using a multivariate statistical approach, J. Trad. Med., 27 (2010), pp. 210–216

95. T Tohge, AR Fernie, Combining genetic diversity, informatics and metabolomics to facilitate annotation of plant gene function, Nature Protocols, 5 (2010), pp. 1210–1227

96. RJM Weber, MR Viant, MI-Pack:Increased confidence of metabolite identification in mass spectra by integrating accurate masses and metabolic pathways, Chemometrics Intel. Lab. Sys., 104 (2010), pp. 75–82

97. K Takemoto, Global architecture of metabolite distributions across species and its formation mechanisms, BioSystems, 100 (2010), pp. 8–13

98. A Acharjee, B Kloosterman, C Maliepaard, Data integration and network reconstruction with omics data using Random Forest regression in potato, Anal. Chim. Acta, 705 (2011), p. 8

99. S Krueger, P Giavalisco, L Krall, MC Steinhauser, D Bussis, B Usadel, UI. Flugge, AR Fernie, L Willmitzer, D teinhauser, A topological map of the compartmentalized Arabidopsis thaliana leaf metabolome, PLoS One, 6 (2011), pp. e17806.1–e1780616

100. KA Aliferis, M Chrysayi-Tokousbalides, Metabolomics in pesticide research and development: review and future perspectives, Metabolomics, 7 (2011), pp. 35–53

101. I Kouskoumvekaki, G Panagiotou, Navigating the Human Metabolome for Biomarker Identification and Design of Pharmaceutical Molecules, J, Biomedicine and Biotechnol (2011) http://dx.doi.org/10.1155/2011/525497

102. GK Kumaraswamy, V Bollina, AC Kushalappa, TM Choo, Y Dion, S Rioux, O Mamer, D Faubert, Metabolomics technology to phenotype resistance in barley against Gibberella zeae, Eur. J. Plant Pathol., 130 (2011), pp. 29–43

103. M Kusano, M Tabuchi, A Fukushima, K Funayama, C Diaz, M Kobayashi, N Hayashi, YN Tsuchiya, H Takahashi, A Kamata, T Yamaya, K Saito, Metabolomics data reveal a crucial role of cytosolic glutamine synthetase 1;1 in coordinating metabolic balance in rice, Plant J., 66 (2011), pp. 456–466

104. N Ohkama-Ohtsu, Y Sasaki-Sekimoto, A Oikawa, Y Jikumaru, S Shinoda, E Inoue, Y Kamide, T Yokoyama, MY Hirai, K Shirasu, Y Kamiya, DJ Oliver, K Saito, 12-oxo-phytodienoic acid-glutathione conjugate is transported into the vacuole in Arabidopsis, Plant Cell Physiol., 52 (2011), pp. 205–209

105. S Osorio, A Bombarely, P Giavalisco, B Usadel, C Stephens, I Araguez, N Medina-Escobar, M Botella, AR Fernie, V Valpuesta, Demethylation of oligogalacturonides by FaPE1 in the fruits of the wild strawberry Fragaria vesca triggers metabolic and transcriptional changes associated with defence and development of the fruit, J. Exp. Botany (2011) http://dx.doi.org/10.1093/jxb-erq465

106. A Scalbert, C Andres-Lacueva, M Arita, P Kroon, C Manach, M Urpi-Sarda, D Wishart, Databases on food phytochemicals and their health-promoting effects, J. Agric. Food Chem., 59 (2011), pp. 4331–4348

107. P Giavalisco, Y Li, A Matthes, A Eckhardt, HM Hubberten, H Hesse, S Segu, J Hummel, K Kohl, L Willmitzer, Elemental formula annotation of polar and lipophilic metabolites using 13C, 15N and 34S isotope labelling, in combination with high-resolution mass spectrometry, Plant J., 68 (2011), pp. 364–376

108. O. Fiehn, DK Barupal, T Kind, Extending biochemical databases by metabolomic surveys, J. Biol. Chem., 286 (2011), pp. 23637–23643

109. K Kai, H Takahashi, H Saga, T Ogawa, S Kanaya, D Ohta, Metabolomic characterization of the possible involvement of a Cytochrome P450, CYP81F4, in the biosynthesis of indolic glucosinolate in Arabidopsis, Plant Biotechnol., 28 (2011), pp. 379–385

110. A Yanuar, A Mun'im, ABA ApLagho, RR Syahdi, M Rahmat, H Suhartanto, Medicinal plants database and three dimensional structure of the chemical compounds from medicinal plants in Indonesia Int. J. Computer Sci. Issue, 8 (2011), pp. 180–183

111. A Katoh, S Fukuda, E Fukusaki, T Hashimoto, T Hayasaki, S Kanaya, H Komura, K Nomoto,, M Shojo, KJ Takeno Systems biology in a commercial quality study of the Japanese Angelica radix: toward an understanding of traditional medicinal plants, Am. J. Chinese Med., 39 (2011), pp. 757–777.

112. Y Kaneko, Y Obata, T Nishino, H Kakeya, Y Miyazaki, T Hayasaka, M Setou, A Furusu, S Kohno, Imaging mass spectrometry analysis reveals an altered lipid distribution pattern in the tubular areas of hyper-IgA murine kidneys, Exp. Mol. Pathol., 91 (2011), pp. 614–621

113. KA Aliferis, M Chrysayi-Tokousbalides, Metabolomics in pesticide research and development: review and future perspectives, Metabolomics, 7 (2011), pp. 35–53

114. H Takahashi, T Morimoto, N Ogasawara, S Kanaya, AMDORAP: non-targeted metabolic profiling based on high-resolution LC-MS, BMC Bioinformatics, 12 (2011), pp. 259.1–259.8

115. T Obata, AR Fernie, The use of metabolomics to dissect plant responses to abiotic stress, Cell. Mol. Life Sci., 69 (2012), pp. 3225–3243

116. MA Sartor, A Ade, Z Wright, D States, GS Omenn, B Athey, A Karnovsky, Metab2MeSH: annotating compounds with medical subject headings, Bioinformatics, 28 (2012), pp. 1408–1410.

117. T Asano, K Kobayashi, E Kashihara, H Sudo, R Sasaki, Y Iijima, K Aoki, D Shibata, K Saito,, M Yamazaki, Suppression of camptothecin biosynthetic genes results in metabolic modification of secondary products in hairy roots of Ophiorrhiza pumila Phytochemistry (2012) (in press)

118. I Ahuja, R Kissen, AM Bones, Phytoalexins in defense against pathogens, Trends Sci., 17 (2012), pp. 73–90.

119. Y Okazaki, K Saito, Recent advances of metabolomics in plant biotechnology Plant Biotechnol. Rep., 6 (2012), pp. 1–15

120. G Marti, M Erb, J Boccard, G Glauser, GR Doyen, N Villard, CA Robert, TC Turlings, S, Rudaz, JL Wolfender, Metabolomics reveals herbivore-induced metabolites of resistance and susceptibility in maize leaves and roots, Plant, Cell & Env., 36 (2012), pp. 621–639

121. LM Liberman, R Sozzani, PN Benfey, Integrative systems biology: an attempt to describe a simple weed, Curr. Opin. Plant Biol., 15 (2012), pp. 162–167

122. B Houshyani, P Kabouw, D Muth, RCH de Vos, RJ Bino, HJ Bouwmeester, Characterization of the natural variation in

Arabidopsis thaiana metabolome by the analysis of metabolic distance, Metabolomics, 8 (2012), pp. S131–S145

123. Y Wahyuni, AR Ballester, Y Tikunov, RCH de Vos, KTB Pelgrom, A Maharijaya, E Sudarmonowati, RJ Bino, AG Bovy, Metabolomics and molecular marker analysis to explore pepper (Capsicum sp.) biodiversity, Metabolomics (2012) http://dx.doi.org/10.1007/s11306-0432-6

124. R Sano, R Ara, N Akimoto, N Sakurai, H Suzuki, Y Fukuzawa, Y Kawamitsu, M Ueno, D, Shibata, Dynamic metabolic changes during fruit maturation in Jatropha curcas L. Plant Biotechnol., 29 (2012), pp. 175–178

125. M Ohtani, Y Nakano, T Usami, T Demura, Comparative metabolome analysis of seed kernels in phorbol ester-containing and phorbol ester-free accessions of Jatropha curcas L. Plant Biotechnol., 29 (2012), pp. 171–174

126. SA Khan, PY Chibon, RC de Vos, BA Schipper, E Walraven, J Beekwilder, T van Dijk, R, Finkers, RG Visser, EW van de Weg, A Bovy, A Cestaro, R Velasco, E Jacobsen, HJ Schouten Genetic analysis of metabolites in apple fruits indicates an mQTL hotspot for phenolic compounds on linkage group 16, J. Exp. Botany, 63 (2012), pp. 2895–2908

127. MMN Alla, A Khedr, MM Serag, AZ Abu-Alnaga, RM Nada, Regulation of metabolomics in Atriplex halimus growth under salt and drought stress, Plant Growth Regul., 67 (2012), pp. 281–304

128. K Haug, RM Salek, P Conesa, J Hastings, P de Matos, M Rijnbeek, T Mahendraker, M, Williams, S Neumann, P Rocca-Serra, E Maguire, A Gonzalez-Beltran, SA Sansone, JL Griffin, C Steinbeck, MetaboLights–an open-access general-purpose repository for metabolomics studies and associated meta-data Nucleic Acid Res (2012) http://dx.doi.org/10.1093/nar/gks1004

129. B Wagele, M Witting, P Schmitt-Kopplin, K Suhre, MassTRIX reloaded: combined analysis and visualization of transcriptome and metabolome data, PlosOne, 7 (2012), p. e39860

130. M Peukert, A Matros, G Lattanzio, S Kaspar, J Abadia, HP Mock ,Spatially resolved analysis of small molecules by matrix-assisted laser desorption/ionization mass spectrometric imaging (MALDI-MSI)., New Phytol., 193 (2012), pp. 806–815

131. AR Ballester, MT Lafuente, RC de Vos, AG Bovy, L Gonzalez-Candelas, Citrus phenylpropanoids and defence against pathogens. Part I: Metabolic profiling in elicited fruits, Food Chemistry, 136 (2012), pp. 178–185
132. IT Jolliffe, Principal Component Analysis (2nd Ed.)Springer-Verlag, New York (2002),
133. S Wold, M Sjostrom, L Eriksson, PLS-regression: a basic tool of chemometrics Chemometrics and Intelligent Laboratory Systems, 58 (2001), pp. 109–130
134. A Boulesteix, K Strimmer, Partial least squares: a versatile tool for the analysis of high-dimensional genomic data, Briefings in Bioinformatics, 8 (2006), pp. 32–44
135. M Barker, W Rayens, Partial least squares for discrimination, J. Chemometrics., 17 (2003), pp. 166–173
136. R Bro, Multi-way calibration. multi-linear pls, Journal of Chemometrics, 10 (1996), pp. 47–61
137. AK Smilde, R Bro, P Geladi, Multi-way analysis with application in the chemical sciencesJohn Wiley & Sons, West Sussex, England (2004)
138. AKSmilde, Commentsonmultilinearpls, JournalofChemometrics, 11 (1997), pp. 367–377,
139. SJ Beers, Jamu: The ancient Indonesian art of healingPeriplus Editions (HK) Ltd., Vermont (2001)
140. S Pramono, Jamu in Indonesia daily life and industryInstitute of Natural Medicine University of Toyama (2007)
141. LK Adnyana, AA Soemardji, Evaluation of Pharmacological efficacy of Jamu medicineInstitute of Natural Medicine University of Toyama (2007)
142. HM Sangat, EAM Zuhud, Damayanti EK, Komus penyakit dan tumbuhan obat Indonesia [Ethnofitomedika]Yayasan Obat Indonesia, Jakarta (2000)
143. Afendi FM, Darusman LK, Hirai A, Altaf-Ul-Amin Md, Takahashi H, Nakamura K & Kanaya, S (2010) System Biology Approach for Elucidating the Relationship between Indonesian Herbal Plants and the Efficacy of Jamu. IEEE International Conference on Data Mining Workshops, ICDM 2010 (December 14–17 2010, University of Technology Sydney, Sydney, Australia)

144. Afendi FM, Sulistiyani, Hirai A, Altaf-Ul-Amin Md, Takahashi H, Nakamura K & Kanaya S (2011) Bootstrapping Jamu Dataset to Examine Assignment Consistency of Plants to Jamu Efficacy. The 2nd International Symposium on Temulawak (May 25–27 2011, Bogor Agricultural University, Bogor, Indonesia)
145. KR Gabriel, The biplot graphic display of matrices with application to principal component analysis, Biometrika, 58 (1971), pp. 453–467
146. IK Adnyana, AA Soemardji,Evaluation of pharmacological efficacy of Jamu medicineUniversity of Toyama Institute of Natural Medicine, Toyama (2007)
147. S Pramono, Jamu in Indonesian daily life and industryUniversity of Toyama Institute of Natural Medicine, Toyama (2007)
148. Afendi FM, Altaf-Ul-Amin Md & Kanaya S (2011) Permutation test in evaluating the significance of plants in PLS-DA model of Jamu ingredients. The 7th Asian Crop Science Association Conference, ACSAC 2011 (September 27–30 2011, Bogor Agricultural University, Bogor, Indonesia)
149. JF Hair, WC Black, BJ Babin, RW Anderson, Multivariate Data Analysis (7th Ed.)Prentice Hall (2010)
150. D Hoffmann, Medical herbalism: the science and practice of herbal medicineHealing Arts Press, Rochester Vermont (2003)
151. JA Duke, MJ Bogenschutz-Godwin, J duCellier, PK Duke, Handbook of medicinal herbs (2nd ed.)CRC Press (2002)
152. PI Good, Permutation, Parametric and Bootstrap Tests of Hypotheses (3rd edition)Springer, New York (2005)

Chapter 8

Genomics and Bioinformatics Resources for Crop Improvement

Keiichi Mochida, and Kazuo Shinozaki

RIKEN Plant Science Center, Yokohama, 230-0045 Japan

ABSTRACT

Recent remarkable innovations in platforms for omics-based research and application development provide crucial resources to promote research in model and applied plant species. A combinatorial approach using multiple omics platforms and integration of their outcomes is now an effective strategy for clarifying molecular systems integral to improving plant productivity. Furthermore, promotion of comparative genomics among model and applied plants allows us to grasp the biological properties of each species and to accelerate gene discovery and functional analyses of genes. Bioinformatics platforms and their associated databases are also essential for the effective design of approaches making the best use of genomic resources, including resource integration. We review recent advances in research platforms and resources in plant omics together with related databases and advances in technology.

INTRODUCTION

Sustainable agricultural production is an urgent issue in response to global climate change and population increase (Brown and Funk 2008, Turner et al. 2009). Furthermore, recent increased demand for biofuel crops has created a new market for agricultural commodities. One potential solution is to increase plant yield by designing plants based on a molecular understanding of gene function and on the regulatory networks involved in stress tolerance, development and growth (Takeda and Matsuoka 2008). Recent progress in plant genomics has allowed us to discover and isolate important genes and to analyze functions that regulate yields and tolerance to environmental stress.

The whole genome sequencing of Arabidopsis thaliana was completed in 2000 (The Arabidopsis Genome Initiative 2000). Subsequently, the National Science Foundation (NSF) Arabidopsis 2010 project in the USA was launched with the stated goal of determining the functions of 25,000 genes of Arabidopsis by 2010 (Somerville and Dangl 2000). Technological advances in each omics research area have become essential resources for the investigation of gene function in association with phenotypic changes. Some of these advances include the development of high-throughput methods for profiling expressions of thousands of genes, for identifying modification events and interactions in the plant proteome, and for measuring the abundance of many metabolites simultaneously. In addition, large-scale collections of bioresources, such as mass-produced mutant lines and clones of full-length cDNAs and their integrative relevant databases, are now available (Brady and Provart 2009, Kuromori et al. 2009, Seki and Shinozaki 2009). The genome sequencing project ofjaponica rice was completed in 2005, and the Rice Annotation Project (RAP), which was orchestrated via 'jamboree-style' annotation meetings, aimed to provide an accurate annotation of the rice genome (International Rice Genome

Sequencing Project 2005, Itoh et al. 2007). In conjunction with the rice genome sequence and its related genomic resources, advanced development of mapping populations and molecular marker resources has allowed researchers to accelerate the isolation of agronomically important quantitative trait loci (QTLs) (Ashikari et al. 2005,Konishi et al. 2006, Ma et al. 2006, Kurakawa et al. 2007, Ma et al. 2007).

The aforementioned recent high-throughput technological advances have provided opportunities to develop collections of sequence-based resources and related resource platforms for specific organisms. A schematic representation of each relevant omics resource is shown in Fig. 1, together with the current status of their availabilities for Arabidopsis, rice and soybean as examples. Each biological element that has been measured comprehensively by a high-throughput method is represented in a corresponding plane in a conceptual model with layers ranging from genome to phenome, a model termed 'omic space' (Toyoda and Wada 2004). Such comprehensive models often provide an excellent starting point for designing experiments, generating hypotheses or conceptualizing based on the integrated knowledge found in the omic space of a particular organism. Furthermore, development of such omic resources and data sets for various species allows the comparison of omic properties among species, which promises to be an efficient way to find collateral evidence for conserved gene functions that might be evolutionarily supported. Bioinformatics platforms have become essential tools for accessing omics data sets for the efficient mining and integration of biologically significant knowledge.

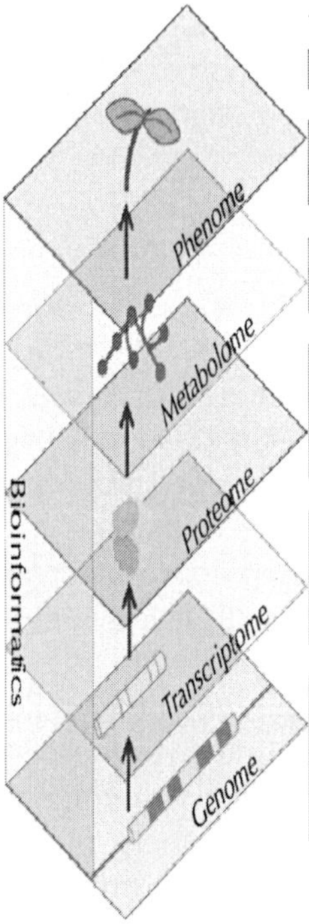

Omics resources	Arabidopsis	Rice	Soybean
Integrated database	TAIR[1]	Gramene[2]	Soy Base[3]
Mutant lincs	Fox line[4] Ac/Ds tag line[5] T-DNA tag line TILLING[3]	Ac/Ds tag line[8], T-DNA tag line9, Tos[17] mutand pancl[10]	TILLING[11]. Ac/Ds tag line[12]
Natural variation	NASC[13] ABSC1[4]	Oryzabase[15], IRRI[16]	Legume basse [17]
Metabolic map	AraCye[18]	RiceCye[19]	(PlantCye[20], MedicCye[21])
Metabolic profile	PRIMe[22], GMD@CSB DB[23]	NA	NA
Protcome / modificome profile	PPDB[24], Phosphat[25]	Rice protcome database [26]	Soybean protcome database[27], Protcomies of oilseeds[28]
Interaction	AtPLN[29], AtPID[30]	NA	NA
Subcellular localization	SUBA[31] NASCprotcome database[32]	Rice protcome database [26]	Soybean Protcome database[27]
Full-length cDNA clones, ESTs	RAFL clones33, RARGE34	KOME[35]	Soybean full-lengthcDNAdatabse[36]
Microarray, GencChip	AtGenExpress[37]Genevesigator[38]	RICEATLAS[39], Genevestigator[38]	Genevestigator[38], SGMD[40], SoyXpress[41]
Noncoding RNA	Arabidopsis MPSS[42]	Rice MPSS[43]	NA
Chip-cip/ Chip seq data.	SIGnAL[44]	NA	NA
Genome sequence, gene annotation	TAIR[1]	RAP-DB[45], TIGR/MSU rice[46]	SoyBase[3], phytozome[47]
Molecular markers, variation data	TAIR[1], Nordborg[48]	Gramene[2], OryzaSNP[49]	Soymap[50]
Re-sequencling	Arabidopsis 1001[51]	NA	NA
Focused gene family database (eg. Transcription factor)	RARTF[52], AGRIS53, DATF[54]	DRTF[55], GRASSIUS[56]	SoybeanTFDB[57], LegumeTFDB[58]

Figure 1: Omic space and related resources in plants. Examples of resources related to each omics instance are represented in Arabidopsis, rice and soybean, as the model plant, as a model monocot and a sequenced crop, and as an important crop recently sequenced, respectively. These resources are accessible from the following URLs or citations. 1. http://www.arabidopsis.org/, 2.http://www.gramene.org/, 3. http://soybase.org/, 4.http://nazunafox.psc.database.riken.jp, 5.http://rarge.gsc.riken.jp/dsmutant/index.pl, 6.http://signal.salk.edu/tabout.html 7. http://tilling.fhcrc.org/, 8.Kolesnik et al. (2004), 9.http://www.postech.ac.kr/life/pfg/risd/, 10.http://tos.nias.affrc.go.jp/, 11.http://www.soybeantilling.org/psearch.jsp, 12.http://mulch.cropsoil.uga .edu/~parrottlab/Mutagenesis/acds/index.php, 13.http://arabidopsis.org.uk/home.html, 14. http://abrc.osu.edu/, 15. http://www.shigen.nig.ac.jp/rice/oryzabase/top/top.jsp, 16.http://www .irri.org/grc/GRChome/home.htm, 17.http://www.legumebase.agr.miyazaki-u.ac.jp/index.jsp, 18.http://www.plantcyc.org:1555/ARA/server.html, 19.http://pathway.gramene.org/gramene/ricecyc.shtml, 20.http://www.plantcyc.org/, 21. http://mediccyc.noble.org/, 22.http://prime.psc.riken.jp/, 23. http://csbdb.mpimp-golm.mpg.de/csbdb/gmd/gmd.html, 24.http://ppdb.tc.cornell.edu/, 25. http://phosphat.mpimp-golm.mpg.de/, 26.http://cdna01.dna.affrc.go.jp/RPD/main_en.html,

27.http://proteome.dc.affrc.go.jp/Soybean/, 28.http://oilseedproteomics .missouri.edu/, 29.http://bioinfo.esalq.usp.br/cgi-bin/atpin.pl, 30.http://atpid. biosino.org/, 31.http://suba.plantenergy.uwa.edu.au/, 32.http://proteomics. arabidopsis.info/, 33. http://www .brc.riken.go.jp/lab/epd/catalog/cdna-clone.html, 34.http://rarge.gsc.riken.jp/, 35.http://cdna01.dna.affrc.go.jp/ cDNA/, 36.http://rsoy.psc.riken.jp/, 37.http://www.arabidopsis.org/portals/ expression/microarray/ATGenExpress.jsp, 38. https://www.genevestigator. com/gv/index.jsp, 39.http://bioinformatics.med.yale.edu/riceatlas/, 40.http:// bioinformatics.towson.edu/SGMD/Default.htm, 41.http://soyxpress.agrenv. mcgill.ca/cgi-bin/soy/soybean.cgi, 42.http://mpss.udel.edu/at/, 43. http:// mpss.udel.edu/rice/, 44.http://signal.salk.edu/, 45. http://rapdb.dna.affrc .go. jp/, 46.http://rice.plantbiology.msu.edu/, 47.http://www.phytozome.net/, 48. http://walnut.usc.edu/, 49.http://www.oryzasnp.org/, 50. http://www.soy-map.org/, 51.http://1001genomes .org/, 52. http://rarge.gsc.riken.jp/rartf/, 53. http://arabidopsis.med.ohio-state.edu/, 54.http://datf.cbi.pku.edu.cn/, 55. http://drtf.cbi.pku.edu.cn/, 56.http://grassius.org/, 57. http://soybeantfdb .psc. riken.jp, 58.http://legumetfdb.psc.riken.jp/.

We provide an overview of several representative resources available for use in omics plant research, with particular emphasis on recent progress related to crop species. We describe sequence-related resources such as whole genome, protein coding and non-coding transcripts, and provide sequencing technology updates. We then review resources for genetic map-based approaches such as QTL analyses and population studies. We also describe the current status of resources and technologies for transcriptomics, proteomics and metabolomics, and review the comprehensive profiling of elements in each omics field as well as instances of their combinatorial uses in the investigation of particular biological systems. Mutant resources for use in phenome research will also be discussed. Finally, the integration of omics data sets across plant species and the progress of comparative genomics will be introduced. Throughout this review, we provide examples of applications of such resources and available databases.

SEQUENCE RESOURCES IN PLANTS

Comprehensively collected sequence data provide essential genomic resources for accelerating molecular understanding of biological properties and for promoting the application of such knowledge. The recent accumulation of nucleotide sequences of model plants, as well as of applied species such as crops and domestic animals, has provided fundamental information for the design of sequence-based research applications in functional genomics. In this section, we describe recently developed plant sequence resources. Species-specific nucleotide sequence collections also provide opportunities to identify the genomic aspects of phenotypic characters based on genome-wide comparative analyses and knowledge of model organisms (Cogburn et al. 2007, Flicek et al. 2008, Paterson 2008, and Tanaka et al. 2008).

Genome Sequencing Projects

The first genome sequence of a plant was completed for A. thaliana, which is now used as a model species in plant molecular biology due to its small size, short generation time and high efficiency of transformation. The Arabidopsis genome sequence project was performed as a cooperative project among scientists in Japan, Europe and the USA (Bevan 1997). The genome sequencing was completed and published in 2000 by the Arabidopsis Genome Initiative (AGI) (The Arabidopsis Genome Initiative 2000). The draft genome sequence of rice, both japonica and indica, an important staple food as well as a model monocotyledon, was published in 2002 (Goff et al. 2002, Yu et al. 2002). Subsequently, the genome sequence of japonica rice was completed and published by the International Rice Genome Sequencing Project in 2005 (International Rice Genome Sequencing Project 2005). To date, several genome sequencing projects involving various plant species have been completed (Table 1).

Table 1: Whole genome sequencing projects in plants

Common name	Latin name	Sequencing group	References	URL				
Dicots								
				Mouse ear cress	*Arabidopsis thaliana*	Consortium (AGI)	The Arabidopsis Genome Initiative (2000)	http://www.arabidopsis.org/
Poplar	*Populus trichocarpa*	JGI	Tuskan et al. (2006)	http://genome.jgi-psf.org/Poptr1_1/Poptr1_1.home.html				
Lyre-leaf rockcress	*Arabidopsis lyrata*	JGI		http://genome.jgi-psf.org/Araly1/Araly1.home.html				
Pink shepherd's purse	*Capsella rubella*	JGI		(http://www.jgi.doe.gov/sequencing/why/3066.html)				
Rapeseed	*Brassica rapa*	Consortium (MGBP)		http://www.brassica-rapa.org/BRGP/index.jsp				
Tomato	*Solanum lycopersicum*	Consortium (ITGSP)		http://solgenomics.net/				
Potato	*Solanum tuberosum*	Consortium (PGSC)		http://www.potatogenome.net/index.php/Main_Page				
Barrel medic	*Medicago truncatula*	Consortium (IMGAG)		http://www.medicago.org/genome/				
	Lotus japonicus	Consortium	Sato et al. (2008)	http://www.kazusa.or.jp/lotus/				
Monkey flower	*Mimulus guttatus*	JGI		(http://www.jgi.doe.gov/sequencing/why/3062.html)				
Soybean	*Glycine max*	JGI	Schmutz et al. (2010)	http://www.phytozome.net/soybean.php				

Cassava	*Manihot esculenta*	JGI		http://www.phytozome.org/cassava.php				
Grape	*Vitis vinifera*	Consortium	Jaillon et al. (2007)	http://www.genoscope.cns.fr/externe/GenomeBrowser/Vitis/				
Columbine	*Aquilegia formosa*	JGI		(http://www.jgi.doe.gov/sequencing/why/51280.html)				
Eucalyptus	*Eucalyptus grandis*	JGI		http://bioinformatics.psb.ugent.be/genomes/view/Eucalyptus-grandis				
Papaya	*Carica papaya*	Consortium	Ming et al. (2008)	http://asgpb.mhpcc.hawaii.edu/papaya/				
Castor bean	*Ricinus communis*	TIGR		http://castorbean.jcvi.org/				
Monocots								
				Rice	*Oryza sativa japonica*	Consortium (IRGSP)	International Rice Genome Sequencing Project (2005)	http://rgp.dna.affrc.go.jp/E/IRGSP/index.html
Rice	*Oryza sativa indica*	Beijing Genomics Institute	Yu et al. (2002)	http://rice.genomics.org.cn/rice/index2.jsp				
Maize	*Zea mays*	Consortium	Schnable et al. (2009)	http://www.maizegdb.org/				
Sorghum	*Sorghum bicolor*	JGI	Paterson et al. (2009)	http://genome.jgi-psf.org/Sorbi1/Sorbi1.home.html				
	Brachypodium distachyon	JGI, Consortium (IBI)	The International Brachypodium Initiative (2010)	http://www.brachypodium.org/				
Wheat	*Triticum aestivum*	Consortium (IWGSC)		http://www.wheatgenome.org/				

Barley	*Hordeum vulgare*	Consortium (IBSC)		http://www.public.iastate.edu/~imagefpc/IBSC%20 Webpage/ IBSC%20 Template-home.html				
Other								
				Moss	*Physcomitrella patens*	JGI	Rensing et al. (2008)	http://genome.jgi-psf.org/Phypa1_1/Phypa1_1.home.html
Gemmiferous spike moss	*Selaginella moellendorffii*	JGI		http://genome.jgi-psf.org/Selmo1/Selmo1.home.html				
	Cyanidioschyzon merolae	Consortium	Matsuzaki et al. (2004)	http://merolae.biol.s.u-tokyo.ac.jp/				

There are a number of providers for plant genome sequences and annotations. Phytozome is a Web-accessible information resource providing genome sequences and annotations of various plant species. This resource is a joint project of the Department of Energy's Joint Genome Institute (DOE–JGI) and the Center for Integrative Genomics, and is intended to facilitate comparative genomic studies among green plants (http://www.phytozome.net/Phytozome_info.php). The current version of Phytozome (ver. 5.0, January 2010) consists of 18 plant species that were sequenced by JGI and other sequencing projects. Gramene (http://www.gramene.org/) is an information resource established as a portal site for grass species, and it provides various kinds of information related to grass genomics, including genome sequences (Ware 2007, Liang et al. 2008). The current version of Gramene (#30, October 2009) provides genome sequence information for 15 plant species, including five wild rice genome assemblies.

According to data provided on the Entrez Genome Project Web site (http://www.ncbi.nlm.nih.gov/sites/entrez?db =genomeprj), as of November 2009, >150 instances of genome projects in species of Viridiplantae have been tracked, including agronomically important crops such as staple foods, fruit trees, medical plants and a number of green alga species. With the ongoing innovations in next-generation sequencing technologies, the release of sequenced genomes is expected to accelerate (Ossowski et al. 2008, Ansorge 2009). Whole-genome sequence information allows us to derive sets of important genomic features, including the identification of protein-coding or non-coding genes and constructs such as gene families, regulatory elements, repetitive sequences, simple sequence repeats (SSRs) and guanine–cytosine (GC) content. These data sets have become primary sequence material for the design of genome sequence-based platforms such as microarrays, tiling arrays or molecular markers, as well as for reference data sets for the integration of omics elements into a genome sequence. Chromosome-scale comparisons identifying conserved similarities of gene coordinates facilitate documentation of segmental and tandem duplications in related species (Haas et al. 2004, De Bodt et al. 2005). Whole-genome comparisons identifying chromosomal duplication and conserved synteny among related species provide evidence for hypotheses on comparative evolutionary histories with regard to the diversification of species in a related lineage (Paterson et al. 2009, Schnable et al. 2009).

Large-Scale Collections of Expressed Sequence Tags (Ests) and Cdna Clones

ESTs are created by partial 'one-pass' sequencing of randomly picked gene transcripts that have been converted into cDNA (Adams et al. 1993). Since cDNA and EST collections can be acquired regardless of genomic complexity, this approach has been applied not only to model species but also to a number of applied species with large genome sizes due to polyploidy and/or to their number of repetitive sequences. As of November 2009, there are >63 million ESTs in the National Center for Biotechnology Information (NCBI)'s dbEST, a public domain EST database (http://www.ncbi.nlm.nih.gov/dbEST/) that includes a number of plant species (Table 2) (Boguski et al. 1993).

Table 2: Numbers of ESTs and unified transcripts in plants (November 2009)

Species	**No. of ESTs (dbEST)**	**No. of entries (UniGene)**
Physcomitrella patens	382,584	18,870
Picea glauca (white spruce)	299,455	22,472
Picea sitchensis (Sitka spruce)	175,662	18,838
Pinus taeda (loblolly pine)	328,628	18,921
Aquilegia formosa×Aquilegia pubescens	85,039	8,046
Arabidopsis thaliana (thale cress)	1,527,298	30,579
Artemisia annua (sweet wormwood)	85,402	9,462
Brassica napus (rape)	643,601	26,733
Brassica oleracea	59,946	5,617
Brassica rapa (field mustard)	44,570	14,497
Capsicum annuum	116,541	8,868
Citrus clementina	118,365	9,123
Citrus sinensis (Valencia orange)	208,909	15,808
Glycine max (soybean)	1,422,604	33,001
Gossypium hirsutum (upland cotton)	268,786	21,738
Gossypium raimondii	63,577	3,297
Helianthus annuus (sunflower)	133,682	12,216
Lactuca sativa (garden lettuce)	80,781	7,940
Lotus japonicus	195,385	14,493

Malus × domestica (apple)	324,308	23,731
Medicago truncatula (barrel medic)	269,237	18,098
Nicotiana tabacum (tobacco)	317,190	24,069
Populus tremula × Populus tremuloides (hybrid aspen)	76,160	9,652
Populus trichocarpa (western balsam poplar)	89,943	14,965
Prunus persica (peach)	79,203	7,620
Raphanus raphanistrum (wild radish)	164,119	18,788
Raphanus sativus (radish)	83,034	17,649
Solanum lycopersicum (tomato)	296,848	18,228
Solanum tuberosum (potato)	236,568	18,784
Theobroma cacao	159,320	24,958
Vigna unguiculata (cowpea)	187,443	15,740
Vitis vinifera (wine grape)	357,856	22,083
Selaginella moellendorffii	93,806	8,810
Hordeum vulgare (barley)	501,614	23,595
Oryza sativa (rice)	1,249,110	40,978
Panicum virgatum (switchgrass)	436,535	20,973
Saccharum officinarum (sugarcane)	246,892	15,594
Sorghum bicolor (sorghum)	209,814	13,899
Triticum aestivum (wheat)	1,067,290	40,349
Zea mays (maize)	2,018,798	97,123
Chlamydomonas reinhardtii	204,076	11,310
Volvox carteri	132,038	5,638

Because EST data collected from the cDNA libraries of a particular organism consist of redundant sequence data derived from the same gene locus or transcription unit, it is often necessary to perform EST grouping by transcription units and to assemble these groups in order to create a consolidated alignment and representative sequence of each transcript before further analysis. Such steps are performed computationally: a typical work flow consists of 'base calling', i.e. converting the output trace of a sequencer to identified nucleotide data, followed by a cleaning step involving identification and removal of contaminated sequences, the masking out of cloning vector sequences, clustering of identical sequences and alignment of clustered sequences (Ewing et al. 1998, Huang and Madan 1999, Masoudi-Nejad et al. 2006). Then, the obtained data sets of representative transcripts can be used as unified transcript data. There are several data resources that provide such unified data sets of plants, such as NCBI-UniGene, PlantGDB, TIGR Plant Gene Index and HarvEST (Feolo et al. 2000, Lee et al. 2005, Close et al. 2007, Duvick et al. 2008).

The comprehensive and rapid accumulation of cDNA clones together with mass volume data sets of their sequence tags have become significant resources for functional genomics. ESTs derived from various kinds of tissues, including tissues from organisms in a range of developmental stages or under stress, could significantly facilitate gene discovery as well as gene structural annotation, large-scale expression analysis, genome-scale intraspecific and interspecific comparative analysis of expressed genes and the design of expressed gene-oriented molecular markers and probes for microarrays (Ogihara et al. 2003, Zhang et al. 2004, Kawaura et al. 2006, Mochida et al. 2006).

Full-Length cDNA Projects

Although partial cDNAs are useful for rapidly creating catalogs of expressed genes, they are not suitable for further study of gene function. This is because the most popular method for preparing a cDNA library does not provide a full-length cDNA that includes the capped site sequence. The biotinylated cap trapper method, which uses trehalose-thermostabilized reverse transcriptase and is an efficient method for constructing full-length cDNA-enriched libraries, was developed by Hayashizaki's group at RIKEN about 10 years ago. Full-length cDNA libraries and large-scale sequence data sets of clones have become invaluable resources for life science projects studying various species (Hayashizaki 2003, Imanishi et al. 2004, Maeda et al. 2006, Tanaka et al. 2008, Yamasaki et al. 2008a).

The sequence resources derived from full-length cDNAs can also help substantially in identifying transcribed regions in completed or draft genome sequences. In Arabidopsis and rice, full-length cDNA sequences have been used to identify genomic structural features such as transcription units, transcription start sites (TSSs) and transcriptional variants (Seki et al. 2002b, Iida et al. 2004, Itoh et al. 2007, Yamamoto et al. 2009). In species for which we have draft genomes, such asPhyscomitrella, soybean and poplar, full-length cDNA clones have been sequenced to help consolidate genomic infrastructure; this should also contribute to gene discovery (Nanjo et al. 2007, Ralph et al. 2008a,Umezawa et al. 2008) (Table 3). Full-length cDNAs are also useful for determining the three-dimensional (3D) structures of proteins by X-ray crystallography and nuclear magnetic resonance (NMR) spectroscopy and for functional biochemical analyses of expressed proteins in the molecular interactions of protein–ligands, protein–proteins and protein–DNAs. Furthermore, recent advances in proteomics infrastructure require comprehensive data sets of the full-length amino acid sequences used to assign peptides to a protein. These advances also necessitate functional annotations to support systematic knowledge mined for proteins corresponding to identified peptides and for residues modified by, for example, phosphorylation, for use in combination with comparative analyses of modificomic events among species. The full-length cDNA library has also contributed importantly to functional analysis by creating overexpressors used in reverse genetics. The advent of function-based gene discovery by systems such as full-length cDNA overexpressor (FOX) gene hunting, which use full-length cDNA transgenic plants as overexpressors, has provided an effective approach to high-throughput discovery of functional genes associated with phenotypic changes (Ichikawa et al. 2006, Fujita et al. 2007, Kondou et al. 2009).

Table 3: Large-scale collections of full-length cDNA clones in plants

Species	Database	References
Arabidopsis thaliana	http://rarge.gsc.riken.jp/	Seki et al. (2002b)
Citrus species		Marques et al. (2009)
Cryptomeria japonica		Futamura et al. (2008)
Glycine max	http://rsoy.psc.riken.jp/	Umezawa et al. (2008)
Hordeum vulgare	http://www.shigen.nig.ac.jp/barley/	Sato et al. (2009)
Manihot esculenta	http://amber.gsc.riken.jp/cassava/	Sakurai et al. (2007)
Oryza rufipogon		Lu et al. (2008)
Oryza sativa, (japonica)	http://cdna01.dna.affrc.go.jp/cDNA/	Kikuchi et al. (2003)
Oryza sativa, (indica)	http://www.ncgr.ac.cn/ricd	Liu et al. (2007)

Physcomitrella patens	http://www.brc.riken.go.jp/lab/epd/catalog/p_patens.html	Nishiyama et al. (2003)
Populus nigra	http://rpop.psc.riken.jp/index.pl	Nanjo et al. (2004); Nanjo et al. (2007)
Populus trichocarpa		Ralph et al. (2008a)
Thellungiella halophila		Taji et al. (2008)
Triticum aestivum	http://trifldb.psc.riken.jp/	Kawaura et al. (2009); Ogihara et al. (2004)
Zea mays	http://www.maizecdna.org/	Soderlund et al. (2009)

Recently, full-length enriched cDNA libraries have been constructed for non-sequenced crops or forestry species, such as wheat (Triticum aestivum), barley (Hordeum vulgare), cassava (Manihot esculenta), Japanese cedar (Cryptomeria japonica) and Sitka spruce (Picea sitchensis), as well as for plant species showing specific biological characters such as salt tolerance in salt cress (Thellungiella halophila) (Table 3). These full-length cDNA libraries have been used to identify biological features through comparisons of target sequences with those of model organisms such as Arabidopsis, rice and poplar. These libraries also serve as primary sequence resources for designing microarray probes and as clone resources for genetic engineering to improve crop efficiency (Sakurai et al. 2007, Futamura et al. 2008, Ralph et al. 2008b, Taji et al. 2008).

Because of the various key functionalities of full-length cDNA resources in omic space, it is also essential to establish relevant information resources that provide gateways to these resources as well as to integrate related data sets derived from other omics fields and species (Sakurai et al. 2005,Mochida et al. 2009b).

Ultrahigh-Throughput DNA Sequencing

During the past decade, the Sanger sequencing method has been used to complete sequencing of microbial and higher eukaryote genomes. In recent years, a number of alternative technologies, which are adaptations of methods such as pyrosequencing procedures, massively parallel DNA sequencing or single molecule sequencing, have become available (Margulies et al. 2005, Ansorge 2009). Such new sequencing technologies have provided us with new opportunities to be addressed at the entire genome level in the fields of comparative genomics, metagenomics and evolutionary genomics (Varshney et al. 2009).

Whole-Genome Resequencing

Next-generation sequencing technology coupled with reference genome sequence data allows us to discover variations among individuals, strains and/or populations. Nucleotide polymorphisms are effectively identified by mapping sequence fragments onto a particular reference genome data set, a capability that is of immense importance in all genetic research. A whole-genome resequencing project to discover whole-genome sequence variations in 1,001 strains (accessions) of Arabidopsis will result in a data set that will become a fundamental resource for promoting future genetics studies to identify alleles in association with phenotypic diversity across the entire genome and across the entire species range (http://1001genomes.org/) (Weigel and Mott 2009). In rice, a high-throughput method for genotyping recombinant populations that used whole-genome resequencing data generated by the Illumina Genome Analyzer was performed (Huang et al. 2009). One of the most anticipated innovations for next-generation sequencers is the application to whole-genome de novo sequencing. Although, to date, this approach has been realized only in bacterial genomes (Farrer et al. 2009, Moran et al. 2009), a number of attempts are being made to realize this advance in higher species.

Comprehensive Discovery of Small RNAS (sRNAS)

In plants, sRNAs, including microRNAs (miRNAs), short interfering RNAs (siRNAs) and trans-acting siRNAs (ta-siRNAs), are also playing roles as crucial components of epigenetic processes and gene networks involved in development and homeostasis (Ruiz-Ferrer and Voinnet 2009). These RNA molecules are important targets that should be comprehensively identified and their expression should be analyzed using next-generational genomic technologies (Nobuta et al. 2007, Chellappan and Jin 2009). In maize, sRNAs in the wild type and in the isogenic mop1-1 loss-of-function mutant were analyzed by deep sequencing using Illumina's sequencing-by-synthesis (SBS) technology to characterize the complement of maize sRNA (Nobuta et al. 2008). In poplar, expressed sRNAs from leaves and vegetative buds were also discovered using high-throughput Roche 454 pyrosequencing, subsequently the genes of miRNA families, including the novel ones, were identified (Barakat et al. 2007). Deep sequencing ofBrachypodium sRNAs at the global genome level has also been performed, resulting in identification of miRNAs involved in the cold stress response (J. Zhang et al. 2009). The plant miRNA database (PMRD) is a useful information resource on plant miRNA and is available on the Web (http://bioinformatics.cau.edu.cn/PMRD/) (Z. Zhang et al. 2009).

RESOURCES FOR VARIATION ANALYSIS

Recent innovations related to DNA sequencing technology and the rapid growth of genome and cDNA sequence resources allow us to design various types of molecular markers covering entire genomes (Feltus et al. 2004). For high-throughput genotyping, a number of platforms have been developed that have been applied to genetic map construction, marker-assisted selection and QTL cloning using multiple segregation populations (Hori et al. 2007). Such genotyping systems have also been used in post-genome sequencing projects such as genotyping of genetic resources, accessions to evaluate population structure and association studies to identify genetic loci involved in phenotypic changes of species. This recent expansion of analysis platforms addressing genome-wide polymorphisms provides an essential resource in the 'variome' study of plants.

Molecular Markers

Accumulation and saturation of available genetic markers contribute to advances in marker-assisted genetic studies and are important resources with a wide range of applications. Genetic markers designed to cover a genome extensively allow not only identification of individual genes associated with complex traits by QTL analysis but also the exploration of genetic diversity with regard to natural variations (Feltus et al. 2004,

Varshney et al. 2005, Caicedo et al. 2007). With the progress of genome sequencing and large-scale EST analysis in various species, these sequence data sets have become quite efficient sequence resources for designing molecular markers. A number of attempts to design polymorphic markers from accumulated sequence data sets have been made for various species. Several genome-wide rice (Oryza sativa) DNA polymorphism data sets have been constructed based on alignment between japonica and indica rice genomes (Han and Xue 2003, Shen et al. 2004). Large-scale EST data sets are also important resources for discovery of sequence polymorphisms, especially for allocating expressed genes onto a genetic map. Therefore, computational discovery of ESTbase single-nucleotide polymorphisms (SNPs) and/or EST-SNP markers for the purpose of identifying sequence-tagged site (STS) markers has progressed for numerous species, including barley, wheat, maize, melon, Brassica, common bean and sunflower (Kota et al. 2001, Kantety et al. 2002, Kota et al. 2003, Torada et al. 2006, Heesacker et al. 2008, Kota et al. 2008, Blair et al. 2009, Deleu et al. 2009, Kaur et al. 2009, Li et al. 2009).

Several databases provide information on molecular markers in plant species. PlantMarkers is a genetic marker database that contains predicted molecular markers, such as SNP, SSR and conserved ortholog set (COS) markers, from various plant species (Heesacker et al. 2008). GrainGenes is a popular site for Triticeae genomics; it also provides genetic markers and linkage map data on wheat, barley, rye and oat (Carollo et al. 2005). Gramene is a database for plant comparative genomics that provides genetic maps of various plant species (Liang et al. 2008). The Triticeae Mapped EST database (TriMEDB) provides information regarding mapped cDNA markers that are related to barley and their wheat homologs (Mochida et al. 2008).

Platforms for Variation Analyses

High-throughput polymorphism analysis is an essential tool for facilitating any genetic map-based approach. So far, genome-wide genotyping using a hybridization-based SNP typing method has been used to analyze representative Arabidopsis ecotypes and rice strains, and the data sets containing the calculated genome-wide variation patterns for each species have been released. As typified by the Arabidopsis 1,001 project, genome-wide variation study is a key analysis that should be performed after genome sequencing has been completed for a particular reference strain. Therefore, the demand for high-throughput and cost-effective platforms for comprehensive variation analysis (also called variome analysis) has rapidly increased.

As we have already mentioned, whole-genome resequencing approaches are already being realized as a direct solution for variome analysis in species whose reference genome sequence data are available. Diversity Array Technology (DArT) is a high-throughput genotyping system that was developed based on a microarray platform (http://www.diversityarrays.com/index .html) (Jaccoud et al. 2001, Wenzl et al. 2007). In various crop species such as wheat, barley and sorghum, DArT markers have been used together with conventional molecular markers to construct denser genetic maps and/or to perform association studies (Crossa et al. 2007, Peleg et al. 2008, and Mace et al. 2009). In barley and wheat, Affymetrix GeneChip Arrays have been used to discover nucleotide polymorphisms as single-feature polymorphisms based on the differential hybridization of GeneChip probes (Rostoks et al. 2005, Bernardo et al. 2009). The Illumina GoldenGate Assay allows the simultaneous analysis of up to 1,536 SNPs in 96 samples and has been used to analyze genotypes of segregation populations in order to construct genetic maps allocating SNP markers in crops such as barley, wheat and soybean (Hyten et al. 2008, Akhunov et al. 2009, and Close et al. 2009).

TRANSCRIPTOME RESOURCES IN PLANTS

Comprehensive and high-throughput analysis of gene expression, called transcriptome analysis, is also a significant approach to screen candidate genes, predict gene function and discover cis-regulatory motifs. The hybridization-based method, such as that used in microarrays and GeneChips, has been well established for acquiring large-scale gene expression profiles for various species. The recent rapid accumulation of data sets containing large-scale gene expression profiles and the ability of related databases to support the availability of such large repositories of data has provided us with access to large amounts of information in the public domain. This public domain data are an efficient and valuable resource for many secondary uses, such as co-expression and comparative analyses. Furthermore, as a next-generation DNA sequencing application, deep sequencing of short fragments of expressed RNAs, including sRNAs, is quickly becoming an efficient tool for use with genome-sequenced species (Harbers and Carninci 2005, de Hoon and Hayashizaki 2008).

Sequence Tag-Based Platforms in Transcriptomics

Large-scale sequencing of ESTs from cDNA libraries was an early approach for acquiring transcriptome profiles. In this approach, ESTs that are randomly sequenced in an unbiased cDNA library are classified into clusters of transcript sequences using sequence-clustering and/or assembling methods. Then, the abundance of transcripts expressed in each tissue is estimated by counting the number of ESTs with identifiers for each cDNA library and/or each sequence cluster. The same methodological principle has been applied in human and mouse in the form of a 'body map' to derive the transcriptome in various organs (Hishiki et al. 2000, Kawamoto et al. 2000, Ogasawara et al. 2006). Moreover, this principle has also been used in the digital differential display (DDD) tool, which is a component of NCBI's UniGene database and has been applied in large-scale cDNA projects for various species, including plants (Mochida et al. 2003, Fei et al. 2004, Sterky et al. 2004, Zhang et al. 2004). Although this approach, coupled with cDNA clone resources, has facilitated gene discovery and expression profiling, its disadvantages include cost and limited resolution due to large-scale sequencing.

Serial analysis of gene expression (SAGE) is a method based on deep sequencing of short read cDNA tags. SAGE allows the identification of a large number of transcripts present in tissues and enables quantitative comparison of transcriptomes (Velculescu et al. 1995). SAGE is designed to generate a short specific tag (13–15 bp) from the 3′ end of each mRNA present in a sample, after which >10 tags are concatenated and cloned to generate a SAGE library. The sequencing of selected clones from the SAGE library allows efficient collection of transcript tag sequences. A data set of genome sequences or large-scale

ESTs is required to identify genes corresponding to each SAGE tag. Some derivatives of the original protocol (MAGE, SADE, microSAGE, miniSAGE, longSAGE, superSAGE, deepSAGE, 5′ SAGE, etc.) have been developed to improve and expand the utility of SAGE (Hashimoto et al. 2004, Anisimov 2008). For example, superSAGE is an improved version of SAGE that produces 26 bp fragment tags from cDNAs. This method has been applied to simultaneous and quantitative gene expression profiling of both host cells and their eukaryotic pathogens in rice (Matsumura et al. 2003). The 26 bp superSAGE tags have also been used to design probes directly for oligo microarrys (Matsumura et al. 2008).

Another sequencing-based technology is massively parallel signature sequencing (MPSS). MPSS uses a unique method to quantify gene expression levels; it generates millions of short sequence tags per library by sequencing 16–20 bp from the 3′ side of cDNA using a microbead array (Brenner et al. 2000). Databases containing MPSS data on plant species, including Arabidopsis, rice, grape and Magnaporthe grisea (the rice blast fungus), are available online (http://mpss.udel.edu) (Nakano et al. 2006). In addition, the MPSS method has also been used to perform genome-scale discovery and expression profiling of sRNAs in Arabidopsis and rice (Lu et al. 2006, Nobuta et al. 2007). The CT-MPSS was a method recently developed for quantitative analysis of the 5′end of transcripts coupled with cap-trapper method for full-length cDNA cloning. This method has been applied to perform high-density mapping of TSS in Arabidopsis to figure out genome-scale instances of plant promoters (Yamamoto et al. 2009). The data set of Arabidopsis CT-MPSS tags is accessible from ppdb (http://www.ppdb.gene.nagoya-u.ac.jp), a plant promoter database that provides promoter annotation of Arabidopsis and rice (Yamamoto and Obokata 2008).

Hybridization-Based Platforms in Transcriptomics

The history of DNA microarray began with a paper from the P. O. Brown laboratory at Stanford University in 1995 (Schena et al. 1995). Since then, microarray- and DNA chip-related technologies have advanced rapidly and their application has expanded to a wide variety of life sciences disciplines. The methodological principle of DNA microarray or GeneChip analysis is to acquire a comprehensive data set of the molecular abundance of each molecule in a given sample based on its simultaneous hybridization with large numbers of DNA molecular species immobilized on a glass slide or on a silicon chip used as a probe set.

DNA microarrays can be classified into two major types: (i) the 'spotting' type, which was developed at Stanford University; and (ii) the 'on-chip synthesis' type based on manufactured probes. The spotting type was widely used during the early years of transcriptome research. This method entailed preparing DNA microarrays by spotting a cDNA solution onto a glass slide. The on-chip (in situ) oligo synthesis-based method is a light-directed chemical synthesis process that combines solid-phase chemical synthesis with photolithographic fabrication techniques. Initially, this method was employed only in conjunction with the Affymetrix-manufactured GeneChip Array system. In the Affymetrix GeneChip system, a known gene or potentially expressed sequence is represented on the chip by 11–20 unique oligomeric probes that are each 25 bases in length. Roche NimbleGen and Agilent Technology offer platforms to manufacture high-density DNA arrays based, respectively, on Roche's proprietary Maskless Array Synthesizer (MAS) technology and on a non-contact industrial inkjet printing process, both of which are also used for in situ oligo synthesis.

With the recent and rapid increase in the number of sequenced species in whole-genome and/or large-scale cDNA clones, a number of DNA microarrays have also been developed for transcriptome analysis in various plant species. For example, Seki and co-workers designed a custom DNA microarray that uses 7,000 full-length cDNA clones of Arabidopsis as probes and then successfully screens genes in response to abiotic stresses using a two-color method (Seki et al. 2002a). With the recent increase in commercially available DNA microarrays, laboratories are able to use a particular DNA microarray design to obtain transcriptome data from many experiments in order to accumulate a more comprehensive resource for organism-specific transcriptome data. AtGenExpress was a multinational effort designed to uncover the transcriptome of A. thaliana. The data sets collected in AtGenExpress have been one of the most comprehensive resources for the Arabidopsis transcriptome to date (Kilian et al. 2007, Goda et al. 2008).

NCBI's Gene Expression Omnibus (GEO) and the European Bioinformatics Institute (EBI)'s ArrayExpress have been serving as the primary archives of transcriptome data in the public domain (Parkinson et al. 2007, Barrett et al. 2009). There are also several more focused databases that provide calculated transcriptome data with user-friendly interfaces and annotations on probes. ATTED II (http://atted.jp/) is a database that provides co-expression analysis data calculated from publicly available Arabidopsis ATH1 GeneChip data (Obayashi et al. 2007, Obayashi et al. 2009). Co-expression analysis data sets generated from comprehensively collected transcriptome data sets have become an efficient resource capable of facilitating the discovery of genes closely correlated in their expression patterns. Genevestigator (https://www.genevestigator.com/gv/index.jsp), which is a reference expression database and meta-analysis system, also provides summary information from hundreds of microarray experiments on various organisms, including Arabidopsis, barley and soybean, with easily interpretable results (Zimmermann et al. 2004). The electronic fluorescent pictograph (eFP) browser provides gene expression patterns collected from Arabidopsis, poplar, Medicago, rice and barley via a user-friendly interface on the Web (http://www.bar .utoronto.ca/) (Winter et al. 2007). The Arabidopsis Gene Expression Database AREX is a database that provides data sets of high-resolution gene expression patterns of root tissues in Arabidopsis (http://www.arexdb.org/index.jsp) (Birnbaum et al. 2003, Brady et al. 2007). The RICEATLAS is a database housing rice transcriptome data covering various types of tissues (http://bioinformatics.med.yale.edu/riceatlas/) (Jiao et al. 2009).

Tiling arrays, which are high-density oligonucleotide probes spanning the entire genome in a particular organism, are a platform for analyzing expressed regions throughout a whole genome; it is an effective method by which to discover novel genes and elucidate their structure. Seki and co-workers performed transcriptome analysis in Arabidopsis under abiotic stress conditions using a whole-genome tiling array and discovered a number of antisense transcripts induced by abiotic stresses (Matsui et al. 2008). The A. thaliana Tiling Array Express (At-TAX) is a whole-genome tiling array resource for developmental expression analysis and transcript identification in Arabidopsis (Laubinger et al. 2008, Zeller et al. 2009). The usefulness of tiling arrays has recently been extended by coupling this platform with the immunoprecipitation method. For example, the binding sites of

AGAMOU-Like15, AGL15, a MADS domain transcriptional regulator promoting somatic embryogenesis, were identified using a chromatin immunoprecipitation (ChIP) approach coupled with the Affymetrix tiling array for Arabidopsis. This method found approximately 2,000 sites (Zheng et al. 2009). Using the methylcytosine immunoprecipitation (mCIP) method in combination with the Arabidopsis tiling array, a comprehensive DNA methylation map of the genome was constructed as an Arabidopsis methylome data set (Zhang et al. 2006). Sequencing of co-precipitated DNAs together with a protein using the next generation sequencer, 'ChIP-seq', has also become an alternative approach (Park 2009).

PLATFORMS AND RESOURCES IN PROTEOMICS

As genome sequencing projects for several organisms have been completed, proteome analysis, which is the detailed investigation of the functions, functional networks and 3D structures of proteins, has gained increasing attention. Large-scale proteome data sets are also an important resource for the better understanding of protein functions in cellular systems, which are controlled by the dynamic properties of proteins. These properties reflect cell and organ states in terms of growth, development and response to environmental changes. The primary objective of functional proteomics was the high-throughput identification of all of the proteins appeared in cells and/or tissues. Recent, rapid technical advances in proteomics (e.g. protein separation and purification methods, advances in mass spectrometry equipment and methodological developments in protein quantification) have allowed us to progress to the second generation of functional proteomics, including quantitative proteomics, subcellular proteomics and various modifications and protein-protein interactions (Rossignol et al. 2006, Jorrin-Novo et al. 2009, Yates et al. 2009). The different Web-accessible plant proteome-related databases are summarized on the proteomics subcommittee of the Multinational Arabidopsis Steering Committee (MASCP) Web site (http://www.masc-proteomics.org/) under the heading of "Proteomic Databases and Resources".

Proteome Profiling

The typical experimental workflow of protein profiling can be summarized as protein sample preparation, separation and detection, then identification. Various technical advances for each step of the process have greatly increased the overall performance of plant proteomics (Jorrin-Novo et al. 2009).

Sample preparation is the most critical step in any proteomics experiment. The method that uses trichloroacetic acid (TCA) and acetone is the most commonly used procedure for protein precipitation. A method using phenol and $NH_4OAC/MeOH$ is also popular for plant tissues. Sample fractionation effectively improves protein detection and increases proteome coverage by reducing sample complexity. Sequential solubilization is an efficient method for fractionating protein samples based on solubility, molecular mass and isoelectric point. By using a series of different reagents to separate proteins by their different solubilities, sequential solubilization is also an effective way to reduce the complexity of proteins in each fraction and to enrich rare proteins (Agrawal et al. 2005).

One-dimensional SDS–PAGE has been widely used to fractionate complex proteins based on their molecular masses. For high-resolution separation of proteins, two-dimensional gel electrophoresis (2-DE), which uses isoelectric focusing (IEF) as the first dimension and SDS–PAGE as the second dimension, is an effective method. Furthermore, the later development of the immobilized pH gradient (IPG)-IEF as the first dimension has improved reproducibility and resolution. The 2-DE methods have been widely used in proteomics in various species (Islam et al. 2004, Mechin et al. 2004, Chen and Harmon 2006), and databases housing 2-DE information have been developed and released [e.g. the Swiss Institute of Bioinformatics' Expasy SWISS-2DPAGE database (http://au.expasy.org/ch2d/) and the Kazusa DNA Research Institute's Cyano2Dbase (http://bacteria .kazusa.or.jp/cyano_legacy/Synechocystis/cyano2D/index .html)]. Chromatography-based separation methods such as gel filtration chromatography, ion exchange chromatography and affinity chromatography are also effective in separating proteins based on their physicochemical properties.

To identify each protein found in a prepared sample, peptide mass fingerprinting has been widely employed. Currently the most efficient method available consists of two steps: (i) enzymatic digestion of separated proteins into peptides; and (ii) accurate mass measurements of peptides using mass spectrometry (MS). In-gel digestion methods have been widely used to separate protein samples by using 2-DE. With its rapid technical advances, MS continues to play an important role in proteomics. MS equipment consists of a source to ionize samples and a mass spectrometer(s) to detect the ionized samples. In proteomics, usually the matrix-assisted laser desorption ionization (MALDI) method or the electrospray ionization (ESI) method is applied to ionize sample peptides. The MALDI method is typically used in combination with time of flight (TOF) MS as MALDI-TOF-MS while the ESI method is usually used in combination with quadrupole (Q) or ion trap (IT) MS. Recently, MS, such as Q-TOF MS, IT-TOF MS or MALDI Q-TOF MS, has become popular. Furthermore, ion fragmentation by collision-induced dissociation (CID) using tandem MS such as Q-TOF MS or by post-source decay (PSD) using MALDI-TOF MS have been applied to determine peptide amino acid sequences. To identify target proteins, obtained peptide mass fingerprint data are searched against a database of theoretically predicted masses of known amino acid sequences (Hirano et al. 2004, Newton et al. 2004).

In addition to conventional gel electrophoresis-based separation, the gel-free separation method is often used, particularly in the 'shotgun proteomics' approach. In the gel-free method, the protein mixture is directly digested into peptides and separated by the multidimensional separation method. The multidimensional separation method is a combination of different online separation methods including multidimensional protein identification technology (MudPIT). The shotgun approach is suitable for the analysis of proteins that are difficult to separate by 2-DE as well as for high-throughput analysis by automated analytical instruments (Yates et al. 2009).

Fourier transform ion cyclotron resonance mass spectrometry (FT-ICR MS) possesses high resolution, high sensitivity, high dynamic range and high mass measurement accuracy. The high resolution and precision of FT-ICR MS allows us to carry out 'top-down proteomics' in which an intact protein mixture is analyzed directly, without separation (Bogdanov and Smith 2005).

Quantitative Proteomics

Comprehensive quantification of each protein's abundance is quite important for a better understanding of the protein dynamics regulated in response to cellular state and environmental changes. A quantitative proteome approach also plays a crucial role in the discovery of key proteomic changes, including expression, interaction and modification, that are associated with genetic variations and/or visible phenotypic changes (Gstaiger and Aebersold 2009).

Difference gel electrophoresis (DIGE) is a popular method for differential display of proteins for quantitative protein comparison. In DIGE, protein samples are labeled with different fluorescent dyes before 2-D electrophoresis, enabling accurate analysis of differences in protein abundance between samples (Rossignol et al. 2006). This method is an effective way to remove gel to gel variation while significantly increasing accuracy and reproducibility. Isotope-coded affinity tags (ICATs), isobaric tags for relative and absolute quantitation (iTRAQ) and stable isotope labeling with amino acids in cell culture (SILAC) are widely used methods for protein differential display using stable isotope labeling (Jorrin-Novo et al. 2009). Using a single MS/MS analysis, corresponding peptides from each sample are differentially detectable based on mass shift caused by the differential isotopes; this allows comparison of the relative abundance of the two samples. Recently, label-free quantitative techniques have been developed to facilitate high-throughput comparisons of proteomic expression. For label-free quantification, the proteomes from each of two samples are separately analyzed using liquid chromatography (LC)-MS/MS. Then, each MS1 spectrum is aligned to calculate relative protein abundance changes based on ion intensity changes such as peptide peak areas or peak heights in chromatography. Finally, MS/MS analysis is used to identify peptides (Gstaiger and Aebersold 2009).

Subcellular Proteomics

Large-scale proteome analysis of cell organelles is essential for understanding the enzymatic inventory of a cell organelle; the compartmentalization of metabolic pathways; cellular logistics such as protein targeting, trafficking and regulation; and proteomic dynamics at the organelle level caused by changes in cellular systems (Andersen and Mann 2006, Chen and Harmon 2006, Baginsky 2009). A number of approaches have been applied to analyze the proteome of organelles or subcellular compartments of plant cells such as chloroplasts, etioplasts, amyloplasts, chromoplasts, mitochondria, vacuoles, plasma membranes, nucleus, peroxisomes, cytosolic ribosome and cell wall (Baginsky 2009). Proteomic analyses of chloroplasts, mitochondria and further fractionations have been carried out to determine detailed localizations of protein in different suborganelle compartments. Techniques for quantitative proteomics, such as the ICAT and iTRAQ methods described above, are also effective for acquiring quantitative data on proteomes in each organelle. In Arabidopsis, rice and alga, differential proteome profiles of plant plasma membranes were monitored to identify those proteins differentially expressed in response to environmental factors such as cold acclimation, salt stress and bacterial elicitor (Benschop et al. 2007, Katz et al. 2007, Cheng et al. 2009, and Minami et al. 2009).

Several databases provide subcellular proteome information. The rice proteome database (http://gene64.dna.affrc.go .jp/RPD/) is a 2-DE image database for rice that contains data from various tissues as well as subcellular compartments (Komatsu 2005). The Nottingham Arabidopsis Stock Centre (NASC) Proteomics database (http://proteomics.arabidopsis .info/) and the SUB-cellular location database for Arabidopsis proteins (SUBA) (http://suba.plantenergy.uwa.edu.au/) provides subcellular proteome analysis data for Arabidopsis (Dunkley et al. 2006). The soybean proteome database (http://proteome .dc.affrc.go.jp/cgi-bin/2d/2d_view_map.cgi) also provides 2-DE data for various tissues as well as for subcellular compartments (Sakata et al. 2009).

Post-Translational Protein Modifications

Comprehensive approaches to investigate various kinds of post-translational protein modifications also play a key role in the current study of proteomics. It (is also called modificome research) aims to identify modified proteins and to elucidate and coordinate the role of each protein functional modification with its associated biological event (Kwon et al. 2006).

Protein phosphorylation is a critical regulatory step in signaling networks and is a widespread protein modification affecting most basic cellular processes in eukaryotic organisms (Schmelzle and White 2006). Advances in MS-based technologies, accompanied by phosphopeptide enrichment techniques, have allowed us to perform high-throughput, large-scale in vivo phosphorylation site mapping. So far, several plant phosphoproteome studies have been reported (Nuhse et al. 2003, Nuhse et al. 2004, de la Fuente van Bentem et al. 2006, Benschop et al. 2007, Nuhse et al. 2007,Sugiyama et al. 2008). For example, the proteome-wide mapping of in vivo phosphorylation sites in Arabidopsis has recently been achieved by using complementary phosphopeptide enrichment techniques coupled with high-accuracy LC-MS/MS with a Finnigan LTQ-Orbitrap (Sugiyama et al. 2008). The Arabidopsis Protein Phosphorylation Site Database (PhosPhAt) provides information on Arabidopsis phosphorylation sites which were identified by MS by different research groups (http://phosphat.mpimp-golm.mpg.de/). The Plant Protein Phosphorylation Database (P3DB) (http://www.p3db.org/), an information resource for plant phosphoproteomes, provides a resource for protein phosphorylation data from multiple plants (Gao et al. 2009).

Ubiquitination of protein is also one of the major post- translational modifications occurring in eukaryotic cells. Protein ubiquitination is a key regulatory mechanism that controls protein abundance, localization and activity. Several large-scale analyses of protein ubiquitination in plants have been reported (Maor et al. 2007, Manzano et al. 2008, Igawa et al. 2009). For example, in Arabidopsis, affinity purification using an anti-ubiquitin antibody and the subsequent use of MS/MS analysis has been performed to identify ubiquitinated proteins (Igawa et al. 2009).

Platform Advances in Structural Proteomics

Large-scale data sets of protein 3D structures are also crucial information resources for elucidating relationships between protein functions and structures or for analyzing molecules in protein complexes. The International Structural Genomics Organization (ISGO,http://www.isgo.org) was formed to facilitate global structural genomics research efforts (Stevens et al. 2001). The key centers for structural genomics have been the RIKEN Structural Genomics/Proteomics Initiative (RSGI) in

Japan, the Protein Structure Initiative (PSI) in the USA and the structural genomics centers of Europe (Yokoyama et al. 2000). International efforts to determine protein structures have contributed to increases in the number of solved protein structures. Thus, the number of solved protein structures appearing in the protein data bank, PDB (http://www.pdb.org/pdb/home/home.do), which is the most popular resource for biomolecule structure data sets, has dramatically increased during the past decade (Kouranov et al. 2006).

The PSI has promoted large-scale attempts to determine the 3D structure of protein folding. The PSI-1 was started as a pilot phase in 2000 with the aim of promoting the strategic development of tools and infrastructures necessary for large-scale determination of protein structures. The PSI-1 centers developed a systematic workflow pipeline encompassing target selection, cloning, expression and purification, followed by crystallization, X-ray crystallography or NMR methods to solve structures. The final step of this pipeline consisted of structure deposition into a database. In 2005, the PSI shifted to its second phase, which was known as PSI-2. The goal of PSI-2's 'production phase' is to solve more challenging structures such as protein complexes and integral membrane proteins (Fox et al. 2008). The RIKEN SGPI has solved >2,700 protein structures, including 33 from Arabidopsis that appear in the PDB (http://www.rsgi.riken.go.jp/rsgi_e/index.html).

Although methodological bottlenecks still exist in structural proteomics, some methodological advances have played an important role in this field. One of the major bottlenecks is the production of soluble and folded proteins. Most centers for structural genomics use Escherichia coli cells for protein production in their automated pipelines as an application of the cell-based method. Cell-free expression systems have also become important as a way to address several limitations of cell-based methods, such as protein quality and quantity as well as throughput issues. The E. coli cell-free system has been applied to amino acid-selective stable isotope labeling of proteins for NMR spectroscopy (Yabuki et al. 1998, Kigawa et al. 1999). The wheat germ embryo cell-free system has also been developed as a eukaryotic cell-free system and has the advantage of producing multidomain proteins (Madin et al. 2000, Endo and Sawasaki 2003, Endo and Sawasaki 2006). The wheat germ cell-free system has since been incorporated into a robotic automation platform (Cell-Free Sciences Co. Ltd.). A comparative study of protein production from 96Arabidopsis open reading frames (ORFs) using cell-based and cell-free systems was reported by the Center for Eukaryotic Structural Genomics (CESG) group in 2005 (Tyler et al. 2005).

The technology and platform of NMR spectroscopy has also played an important role in structural proteomics. The cell-free systems of E. coliand wheat germ embryo in combination with selected amino acid labeling, as described above, have produced synergistic advances in the promotion of automated protein structure determination using solution NMR methods (Yokoyama 2003). Furthermore, high-resolution multidimensional solid state NMR methods used in combination with cross polarization (CP), magic angle spinning (MAS) and dipolar decoupling (DD) are also becoming the methods of choice for structural analysis of membrane proteins by NMR platforms (Castellani et al. 2002, McDermott 2009). Furthermore, recent hardware improvements in NMR probes, including the cryoprobe for improved sensitivity, the micro-coil probe for sample mitigation and the flow-probe designed to shorten preparation time, provide us with the opportunity to use NMR methods to screen ligands that bind to a particular protein.

X-ray crystallography has been used to determine the protein structures of almost 90% of the protein entries in the PDB (http://www.rcsb.org/pdb/static.do?p=general_information/pdb_statistics/index.html). In particular, the beamlines of third-generation X-ray synchrotrons have become an essential infrastructure for macromolecular crystallography (MX), which is used to determine the 3D structures of macromolecules (such as large proteins and protein complexes) (Samatey et al. 2001). For example, the synchrotron SPring-8 of RIKEN in Japan is used to determine the structures of important membrane proteins and protein complex supermolecules such as Ca^{2+}-ATPase, rhodopsin and flagellin (Palczewski et al. 2000, Samatey et al. 2001, Toyoshima et al. 2003).

By using existing analytical platforms for structural proteomics, the structures of many representative DNA-binding domains (DBDs), namely AP2/ERF, NAC, WRKY, B3 and SBP, of plant-specific transcription factor (TF) families have thus far been determined (Yamasaki et al. 2004, Yamasaki et al. 2008b).

Information Resources in Structural Proteomics

Bioinformatics and related databases are also essential tools for advancing the study of structural proteomics. The methods of computational prediction of protein 3D structure are mainly classified into two methods: template-based modeling (TBM) and free modeling (FM) (Zhang 2008, Zhang 2009b). Free modeling, which is also called 'ab initio' or 'de novo' modeling, is used to predict the 3D structure of proteins, without any information on previously solved structures. A number of Web server and computational tools for free and/or template-based modeling have recently been made available; for example, the I-TASSER internet service, which is used in Critical Assessment of Techniques for Protein Structure Prediction (CASP), was released (Zhang 2009a). Template-based modeling method is a comparative method for matching proteins using evolutionarily related proteins of known structure as a template. There are many Web services and tools (e.g. Swiss Institute of Bioinformatics' SWISS-MODEL server) to support template-based modeling (Schwede et al. 2003). Databases housing previously predicted structures from amino acid sequences by template-based modeling for a wide range of species also exist: the Genomes TO Protein structures and functions (GTOP) database (http://spock.genes.nig.ac.jp/~genome/gtop .html) provides information on protein structures and functions obtained through the application of various computational tools for structure prediction and annotation from the amino acid sequences deduced from annotated genes in sequenced genomes (Fukuchi et al. 2009).

The database for structure-based protein classification, as typified by CATH (http://www.cathdb.info/) and the Structural Classification of Proteins (SCOP) database (http://scop.mrc-lmb.cam.ac.uk/scop/), has provided important clues to the relationships between protein structures, protein functions and protein evolution (Greene et al. 2007, Andreeva et al. 2008). Databases of protein families based on conserved protein domains, such as Pfam, Superfamily and Protein ANalysis THrough Evolutionary Relationships (PANTHER), are important resources for classifying proteins into families. These types of databases are often used for the functional prediction and classification of proteins; for example, such resources are used in genome-wide identification of genes putatively encoding specific TFs (Mi et al. 2005, Wilson et al. 2007, and Finn et al. 2008). Many of these protein family databases can be simultaneously searched using the EBI's Interpro (Mulder and Apweiler 2008, Hunter et al. 2009).

PLATFORMS AND RESOURCES IN METABOLOMICS

Metabolomics aims to understand metabolic systems based on comprehensive and integrated approaches by taking advantage of various measurement instruments to characterize metabolites. Metabolomic approaches allow us to conduct parallel assessments of multiple metabolites and to undertake quantitative analysis of particular metabolites in ways that provide major advantages over chemical-level phenotyping and diagnostic analysis. It is notable that the plant metabolome represents an enormous chemical diversity due to the complex set of metabolites produced in each plant species (Bino et al. 2004, von Roepenack-Lahaye et al. 2004). Therefore, plant metabolomics is not only a great analytical challenge, but is also quite important. In its ability to elucidate plant cellular systems, metabolomics permits us to engineer molecular breeding to improve the productivity and functionality of plants in areas such as stress tolerance, pharmaceutical production, functional foods, biomaterials and energy (Trethewey 2004, Oksman-Caldentey and Saito 2005, Fernie and Schauer 2009). In this section, we introduce metabolomic analytical platforms for plants, metabolic profiling, and their applications in combination with other omics. We also describe plant metabolomics-related computational tools and databases.

Instruments for Metabolomics

Many remarkable technological advances have recently been made in instrumentation related to metabolomics. Metabolomics experiments start with the acquisition of metabolic fingerprints using various analytical instruments such as GC-MS, LC-MS, FT-MS, FT-IR and NMR (Fiehn et al. 2000, Roessner et al. 2001, Fernie et al. 2004).

Methods for sample separation, such as gas chromatography (GC), high-performance or ultraperformance LC and capillary electrophoresis (CE), are typically used in conjunction with various types of MS, as detailed below. CE-MS is an especially effective, high-sensitivity method for separating and analyzing polarized molecules in samples (Ramautar et al. 2009). QMS and TOF MS are also well regarded for use in metabolomics. Triple Q (QqQ) MS (a tandem-type MS) and Q-TOF (a hybrid-type MS) are also used. Methods that do not involve pre-separation of samples, e.g. FT-ICR MS, are also being used, allowing for MSn analysis (Werner et al. 2008).

NMR-based methods are also used in metabolomic analysis (Dixon et al. 2006, Schripsema 2009). These methods can be broadly classified into solution NMR and insoluble or solid-state NMR, according to sample solubility. Using high-resolution (hr)-MAS techniques, it is possible to acquire metabolic fingerprints from insoluble samples and solid-state samples (Bertocchi and Paci 2008). In one-dimensional NMR, protons (^{1}H) are usually observed (^{1}H-NMR) due to the sensitivity and common occurrence of this magnetic nucleus. More detailed analyses, such as metabolite identification or flux analysis, can be obtained with other nuclei, particularly ^{13}C and ^{15}N that are coupled with ^{1}H nuclei in two-dimensional or multidimensional NMR analysis (Kikuchi et al. 2004, Sekiyama and Kikuchi 2007).

The resulting fingerprints, MS or NMR spectra, are pre- processed, including background noise suppression, peak alignment and peak picking. Pre-processed data sets are subsequently used to identify metabolites corresponding to each spectrum signal by searching against compound databases. In non-target analyses, spectrum data sets that include spectra of unknown compounds are subjected to statistical analyses, such as multivariate analysis, to mine the data for biological significance (Tikunov et al. 2005). In target analyses, spectrum data sets that are associated with particular compounds are used as metabolic profiles for each compound in further analyses.

Data analysis is important in the determination of biological significance in metabolomics. Statistical analyses using multivariate analysis, such as principal component analysis (PCA), hierarchical clustering analysis (HCA) and self-organization mapping (SOM), are typically used to classify samples and/or metabolites (Kose et al. 2001, Hirai et al. 2004, Jonsson et al. 2004, Matsuda et al. 2009). The visualization of metabolic profiles on metabolic pathway maps is also often used and is combined with other omics methods, including gene expression profiles of genes encoding enzymes involved in particular pathways (Thimm et al. 2004, Tokimatsu et al. 2005).

Metabolite Profiling In Plants

The systematic collection of metabolite profiles is the initial step in metabolomics. This step can be performed with various instruments capable of high throughput and simultaneous measurement, as we mentioned above. Comprehensive metabolic profile data sets can contribute to the understanding of the cellular system in response to changes in intracellular and extracellular environments. Furthermore, the changes in metabolic profiles associated with genetic variations can be evaluated as chemical phenotypes to identify genes involved in particular metabolic pathways.

A number of studies of metabolic profiling in plant species have been performed that have resulted in the publication of related databases. For example, in the case of Arabidopsis, an NSF-funded multi-institutional project aimed at development of the metabolomics database, Plantmetabolomics, has recently been undertaken (http://lab.bcb.iastate.edu/sandbox/pbais05/alpha/plantmetabolomics_trimmed/index.php). Several databases for Solanaceae species are already available. The Metabolome Tomato Database (MoTo DB) was developed as an LC-MS-based metabolome database (http://appliedbioinformatics.wur.nl/moto/) (Moco et al. 2006). The KOMICS (Kazusa-omics) database collects annotations of metabolite peaks detected by LC-FT-ICR-MS and contains a representative metabolome data set for the tomato cultivar, Micro-tom (Iijima et al. 2008). The Armec Repository Project provides metabolome data on the potato and serves as a data repository for metabolite peaks detected by ESI-MS (http://www.armec.org/MetaboliteLibrary/index.jsp). The Golm Metabolome Database (GMD) provides public access to custom mass spectra libraries and metabolite profiling experiments as well as to additional information and related tools (http://csbdb.mpimp-golm.mpg .de/csbdb/gmd/gmd.html) (Kopka et al. 2005). The MS/MS spectral tag (MS2T) libraries at the Platform for Riken Metabolomics (PRIMe) website provides access to libraries of phytochemical LC-MS2 spectra obtained from various plant species by using an automatic MS2 acquisition function of LC-ESI-Q-TOF/MS (http://prime.psc.riken.jp/lcms/ms2tview/ms2tview.html) (Matsuda et al. 2009). These databases play crucial roles as information resources and repositories of large-scale data sets and also serve as tools for further integration of metabolic profiles containing comprehensive data acquired from other omics research (Akiyama et al. 2008).

Combinatorial Approaches In Metabolomics and Other Omics Resources

Metabolome approaches also support the understanding of global relationships among cellular metabolic systems in combination with other omics instances such as profiles of the transcriptome and proteome, and also genetic variations. So far, these combinatorial approaches have been successfully demonstrated in the well-studied Arabidopsis by taking advantage of the many other omics resources that currently exist, including the whole-genome sequence with mature annotations, large-scale transcriptome data sets and related co-expression data, and bioresources such as collections of mutants and full-length cDNA clones. A conceptual scheme for systematic elucidation of gene-to-metabolites molecular networks through a combinatorial approach using transcriptome and metabolome resources has been demonstrated by Saito's group in the RIKEN Plant Science Center (Saito et al. 2008). Data sets containing transcriptome and metabolome changes of Arabidopsis under stress conditions induced by deficiency of sulfur and nitrogen were analyzed using a batch-learning, self-organizing map (BL-SOM) analysis, enabling identification of genes involved in glucosinolate biosynthesis (Hirai et al. 2004). An integrated approach that comprised metabolome and transcriptome analysis was conducted for investigation of an activation-tagged mutant and overexpressors of an MYB TF, PAP1 gene in order to identify genes involved in anthocyanin biosynthesis in Arabidopsis (Tohge et al. 2005). Co-expression data of the Arabidopsis transcriptome provided by the ATTED-II database have been applied to the investigation of key genes involved in specific metabolic pathways and then to the configuration of a metabolome analysis coupled with mutant lines of the targeted genes (Obayashi et al. 2009). The ATTED-II database was used to identify novel genes involved in lipid metabolism, leading to identification of a novel gene, UDP-glucose pyrophosphorylase3 (UGP3) that is required for the first step of sulfolipid biosynthesis (Okazaki et al. 2009). Co-expression analysis was also used to identify all of the genes related to flavonoid biosynthesis, which led to further detailed analysis of two flavonoid pathway genes UGT78D3 and RHM1 (Yonekura-Sakakibara et al. 2008).

Approaches that integrated metabolome and transcriptome data have also elucidated regulatory networks that act in response to environmental stresses in plants. The metabolic pathways that act in response to cold and dehydration conditions in Arabidopsis were investigated by metabolome analysis using various types of MS coupled with microarray analysis of overexpressors of genes encoding two TFs, DREB1A/CBF3 and DREB2A (Maruyama et al. 2009). Metabolomic profiling was also used to investigate chemical phenotypic changes between wild-type Arabidopsis and a knockout mutant of the NCED3 gene under dehydration stress conditions. The metabolic data were then integrated with transcriptome data to reveal ABA-dependent regulatory networks (Urano et al. 2009).

Metabolome profiling has also been used to evaluate chemical phenotypes of natural variations and/or segregation populations simultaneously. A comprehensive exploration of the association between metabolic and genomic diversity will enable the discovery of key genes involved in metabolic changes and would also aid in the identification of genetic associations between metabolic and/or visible phenotypes (Schauer et al. 2008, Fu et al. 2009). Metabolite QTL (mQTL) analysis using segregated populations has been applied to various plant species such as Arabidopsis, poplar and tomato in a popular forward genetics approach (Morreel et al. 2006, Schauer et al. 2006, Lisec et al. 2008, Rowe et al. 2008; Schauer et al. 2008). Furthermore, along with the recent availability of data sets of genome-wide variation acquired by high-throughput genotyping methods including resequencing, interest in the discovery of the genetic association between nucleotide variation and phenotypic changes has also increased, especially with regard to the identification of key genes that play significant roles in evolutionary histories. The attempts to mine correlative patterns between metabolic and genomic diversities have recently been applied to sesame and rice using seed stocks of natural variations (Laurentin et al. 2008, Mochida et al. 2009a).

Information Resources for Metabolomics

Various information resources related to metabolomics have played crucial roles not only in metabolome research but also in synergistic integration with other omics data. The Web site of metabolome resources at TAIR (http://www.arabidopsis.org/portals/metabolome/index.jsp) provides a summarized list of Web hyperlinks to resources that facilitate metabolome research. In addition, a data set of biological pathway maps is available via the Kyoto Encyclopedia of Genes and Genomes (KEGG) by using a popular database for information on life sciences called the KEGG PATHWAY Database (http://www .genome.jp/kegg/pathway.html) (Kanehisa and Goto 2000, Kanehisa et al. 2008). The Plant Metabolic Network (PMN) is a collaborative project that aims to build plant metabolic pathway databases (http://www.plantcyc.org/). One of its main components, PlantCyc, is a comprehensive plant biochemical pathway database that contains curated information from the literature and from computational analyses of genes, enzymes, compounds, reactions and pathways involved in primary and secondary plant metabolism (http://www.plantcyc.org:1555/PLANT/server.html) that can be visualized using a pathway tool (http://bioinformatics.ai.sri.com/ptools/). AraCyc and PoplarCyc are also available at the PMN Web site and provide manually curated or reviewed information about metabolic pathways in Arabidopsis and poplar, respectively (Mueller et al. 2003). There are also metabolic pathway databases for several other plant species generated by PMN collaborators. MapMan is a tool to project omics datasets including gene expression data onto diagrams of metabolic pathways or other processes (http://mapman.gabipd.org/web/guest) (Thimm et al. 2004). KaPPA-View is another Web-based analysis tool that can be used to superimpose transcriptome and metabolome data onto plant metabolic pathway maps (http://kpv.kazusa .or.jp/kappa-view/) (Tokimatsu et al. 2005). PRIMe is a Web-based service that provides data sets of metabolites measured by multidimensional NMR spectroscopy, GC-MS, LC-MS and CE-MS together with analytical tools to promote integrated approaches using comprehensive data sets within the metabolome and transcriptome (http://prime.psc.riken.jp/) (Akiyama et al. 2008).

MUTANT RESOURCES FOR PHENOME ANALYSIS

Analysis of mutants is an effective approach for investigation of gene function (Springer 2000, Stanford et al. 2001). Comprehensive collections of mutant lines are also essential bioresources for radically accelerating forward and reverse genetics. The available mutant resources for phenome analysis in plant species have been well described in a recent review byKuromori et al. (2009). As described above, various analytical platforms have rapidly evolved, allowing us to discover genes involved in particular phenotypic changes. Along with these analysis platforms, the demands for comprehensive collections of mutants and related information resources have dramatically increased, encouraging high-throughput and genome-wide phenome analysis in plant species (Alonso and Ecker 2006).

Insertion Mutant

With the completion of genome sequencing in plants, insertion mutant resources with index data that document the inserted genomic position have become extremely beneficial resources by which to promote functional analysis of annotated genes that are disrupted by a reverse genetics approach. Transferred DNA-tagged (T-DNA-tagged) lines and transposon-tagged lines have become popular resources for the investigation of insertion mutants in plants. T-DNA-tagged lines have emerged as a popular mutant resource due to the rapid generation of large-scale populations in Arabidopsis (Krysan et al. 1999). The two-component maize transposon, Activator (Ac) / Dissociation (Ds), is a popular system for inducing transposon-based insertion that enables the generation of mutants with a high proportion of single-copy insertions (Long et al. 1993). In rice, the endogenous retrotransposon Tos17, which is activated in particular conditions, is also available for the study of the insertion mutant lines of a japonica rice cultivar, Nipponbare (Miyao et al. 2003, Miyao et al. 2007), and the Web resource that provides information on the rice Tos17 mutant panel with flanking sequences of insertion is available at http://tos.nias.affrc.go.jp/index.html.en. Additionally, the maize Enhancer/Suppressor Mutator (En/Spm) element has also been used as an effective tool for the study of functional genomics in plants (Kumar et al. 2005). The enhancer trap (ET) and the gene trap (GT) constructs have been coupled with T-DNA and Ac/Ds transposons, which additionally facilitates entrapment of genes in monitoring of adjacent promoter or enhancer activity (Sundaresan et al. 1995, An et al. 2005). OryGenesDB (http://orygenesdb.cirad.fr/) is a database that integrates information of available insertion mutant resources in rice (Droc et al. 2009). There are a number of resources for insertion mutant populations with insertion site index-tagged data available for various plant species (Kuromori et al. 2009)

Activation Tagging

Activation tagging (AT) is a popular method for generating gain-of-function mutant populations. The method uses T-DNA or a transposable element containing cauliflower mosaic virus 35S enhancer multimers (Weigel et al. 2000). With transcriptional activation of genes near the insertion, novel phenotypes are expected to appear that will identify genes that are redundant or essential for survival. Mutant resources have then been used to isolate genes from Arabidopsis, rice, petunia and tomato (Kakimoto 1996, Zubko et al. 2002, Mathews et al. 2003, Mori et al. 2007). Recently, AT systems using a transposon of maize En/Spm orAc/Ds have been developed in Arabidopsis and rice, respectively (Schneider et al. 2005, Qu et al. 2008). A number of AT projects have been performed in various plant species such as Arabidopsis, rice and soybean (Weigel et al. 2000, An et al. 2005, Kuromori et al. 2009).

The FOX Hunting System

The FOX gene hunting system is a recently developed novel gain-of-function system that combines a transformation algorithm with large-scale resources for full-length cDNA clones (Ichikawa et al. 2006). A normalized full-length cDNA library of Arabidopsis was first used to generate the Arabidopsis FOX lines, which are available athttp://nazunafox.psc.database.riken.jp. The system was also used to screen salt stress-resistant lines in the T_1 generation produced by the transformation of 43 focused stress-inducible TFs of Arabidopsis (Fujita et al. 2007). Then, the system was applied to a set of full-length rice cDNA clones aiming for in planta high-throughput screening of rice functional genes, with Arabidopsis as the host species (http://ricefox.psc.riken.jp) (Kondou et al. 2009). The FOX lines with rice full-length cDNA transformed into rice plants has been also generated as a gain-of-function mutant resource of rice (Nakamura et al. 2007). A similar technique (in terms of overexpressors using cDNA libraries) has also been carried out in tobacco (Lein et al. 2008).

Chemical and Physical Mutagenesis

Chemical mutagenic agents, such as ethyl methanesulfonate (EMS), sodium azide and methylnitrosourea (MNU), and physical mutagens, such as fast-neutrons, gamma rays and ion-beam irradiation, have been used to generate mutant populations for many years for forward genetics in various plant species. Targeting induced local lesions in genomes (TILLING) was developed as a general reverse-genetic strategy that provides an allelic series of induced point mutations in genes of interest (Till et al. 2004, Till et al. 2006). Because high-throughput TILLING permits the rapid and low-cost discovery of induced point mutations in populations of chemically mutagenized individuals, the method has been applied to various animal and plant species. The TILLING technology can also be used to explore allelic variations that are appeared in natural variation; this technology is called EcoTILLING (Comai et al. 2004, Wang et al. 2006). Several laboratory sites have established TILLING and/or EcoTILLING centers for communities of users as a public service (Barkley and Wang 2008). TILLING projects in rice, tomato and Arabidopsis have been performed at the University of California Davis Genome Center (http://tilling.ucdavis.edu/index.php/Main_Page). A soybean mutation database provides soybean mutagenized lines together with data on TILLed genes and their phenotypes (http://www.soybeantilling.org/index.jsp). RevGenUK at the John Innes Center provides TILLING service for TILLING populations of Medicago truncatula, Lotus japonicus and Brassica rapa(http://revgenuk.jic.ac.uk/about.htm). UTILLdb of INRA is another database for TILLING populations of pea and tomato that provides an interface to search for TILLed lines based on phenotypes (http://urgv.evry.inra.fr/UTILLdb).

Gene Silencing Technologies

Although insertion mutagenesis is an effective method for generating loss-of-function mutants, it also has limitations in the case of redundant genes and lethal mutants. To overcome these limitations, methods to interrupt gene expression have been developed and applied to the functional analysis of plant genes.

RNA interference (RNAi) is a popular method for RNA- mediated gene silencing by sequence-specific degradation of homologous mRNA triggered by double-stranded RNA (dsRNA), which is also known as post-transcriptional gene silencing (PTGS) (Waterhouse et al. 1998, Chuang and Meyerowitz 2000). Constitutive expression of an intron-containing self-complementary hairpin RNA (ihpRNA) has been an effective method for silencing target genes in plants. With demands for conditional silencing of target genes (the silencing of which results in prevention of plant regeneration or embryonic lethality), conditional RNAi systems using a chemical-inducible Cre/loxP recombination system or a promoter of heat shock-inducible genes have been recently developed (Guo et al. 2003, Masclaux et al. 2004). In Arabidopsis, the Complete Arabidopsis Transcriptome MicroArray (CATMA) project has been started to design and produce high-quality gene-specific sequence tags (GSTs) covering most Arabidopsis genes (http://www .catma.org/). Using the GST data set of the CATMA project, the Arabidopsis Genomic RNAi Knock-out Line Analysis (AGRIKOLA) project has also been started with the goal of systematically analyzing Arabidopsis genes by RNAi-based technology (http://www.agrikola.org/index.php?o=/agrikola/html/index) (Hilson et al. 2003). The M. truncatula RNAi Database is also available on the Web as an information resource for RNAi-based gene silencing (https://mtrnai.msi.umn.edu/). Virus-induced gene silencing (VIGS) is a derivative method that takes advantage of the plant RNAi-mediated antiviral defense mechanism. The VIGS system was used to assess the function of almost 5000 random Nicotiana benthamiana cDNAs in disease resistance (Lu et al. 2003a, Lu et al. 2003b).

The chimeric repressor silencing technology (CRES-T) system was developed as a novel method for gene silencing; it takes advantage of the fact that TFs fused to the EAR motif, a plant-specific repression domain, act as dominant repressors in transgenic plants and therefore suppress the expression of target genes (Hiratsu et al. 2003). The CRES-T system has been applied to the TFs annotated for Arabidopsis in order to analyze their biological function and to obtain transgenic plants with agronomically preferable traits. An associated database, FioreDB, is available at http://www.cres-t.org/fiore/public_db/index.shtml (Mitsuda and Ohme-Takagi 2009).

PLANT COMPARATIVE GENOMICS AND DATABASES

The recent accumulation of nucleotide sequences for agricultural species, including crops and domestic animals, now allows us to perform genome-wide comparative analyses of model organisms with the goal of discovering key genes involved in phenotypic characteristics (Sato and Tabata 2006, Itoh et al. 2007, Neale and Ingvarsson 2008). The integration of genomic resources derived from various related species, such as large-scale collections of cDNAs and data from whole-genome sequencing projects, should facilitate sharing of information about gene function between models and applied organisms. This will also accelerate molecular elucidation of cellular systems related to agronomically important traits. A number of information resources for plant genomics accessible on the Web have appeared, along with appropriate analytical tools. Here we highlight integrative databases promoting plant comparative genomics that we have not described previously. The URLs of each integrative database in plant genomics are shown in Table 4.

Table 4: Integrative databases in plants

Database name	Species	URL
TAIR	Arabidopsis	http://www.arabidopsis.org/
SIGnAL	Arabidopsis	http://signal.salk.edu/
RARGE	Arabidopsis	http://rarge.psc.riken.jp/
Rice Genome Annotation Project	Rice	http://rice.plantbiology.msu.edu/
RAP-DB	Rice	http://rapdb.dna.affrc.go.jp/

SOL genomics network	Solanaceae	http://solgenomics.net/
Gramene	Gramineae	http://www.gramene.org/
GrainGenes	Triticeae and Avena	http://wheat.pw.usda.gov/GG2/index.shtml
SoyBase	Soybean	http://www.soybase.org/
MazieGDB	Maize	http://www.maizegdb.org/
CyanoBase	Cyanobacteria	http://genome.kazusa.or.jp/cyanobase/
GDR (Genome Database for Rosaceae)	Rosaceae	http://www.bioinfo.wsu.edu/gdr/
Brassica Genome Gateway	Brassica	http://brassica.bbsrc.ac.uk/
Cucurbit Genomics Database	Cucurbitaceae	http://www.icugi.org/
Phytozome	Plant species (whole genome data available)	http://www.phytozome.net/
PlantGDB	Plant species (whole genome and/or large-scale EST data available)	http://www.plantgdb.org/
EnsemblPlants	Plant species (whole genome data available)	http://plants.ensembl.org/index.html
ChloroplastDB	Plant species (Chloroplast genome data available)	http://chloroplast.cbio.psu.edu/
KEGG PLANT	Plant species (whole genome and/or large-scale EST data available)	http://www.genome.jp/kegg/plant/

Portal Information Resources in Plants

TAIR is one of the most popular and integrated information resource in plant science, and it plays an essential role as a portal site in Arabidopsis research (http://www.arabidopsis .org/) (Swarbreck et al. 2008). The Salk Institute Genomic Analysis Laboratory (SIGnAL) is also an information resource that integrates various data sets of significant omics results mainly related to Arabidopsis (http://signal.salk.edu/). The RIKEN Arabidopsis Genome Encyclopedia (RARGE) provides information on various genomic resources built at RIKEN for

Arabidopsis research (http://rarge.gsc.riken.jp/db_home.pl) (Sakurai et al. 2005). Such portal sites have provided gateways for access to comprehensive omics data and/or bioresources. These sites also house cross-referenced data sets built between each annotated gene and its associated instances, such as gene–full-length cDNA clones, gene–mutants, gene–expression patterns and gene–homologous genes. Therefore, to visualize an annotated gene along with genome sequences and associated information, genome browsers such as Gbrowse have been implemented on Web sites (Donlin 2007).

Gramene is a popular portal site that is not only an integrated rice information resource but also a portal for promoting plant comparative genomics (http://www.gramene.org/). Gramene offers integrated genome-oriented data including gene annotation and molecular markers, and also a QTL database mainly for Gramineae species (Liang et al. 2008).

Along with the launch of genome sequencing projects, portal sites to share the progression of outcomes and to integrate related resources have appeared for various species. The Sol genomics network is a portal site for Solanaceae genome resources that includes information on the tomato genome sequencing project (http://solgenomics.net/) (Mueller et al. 2005). SoyBase is a resource portal site for genomic soybean research, and it includes released whole-genome sequence data (http://soybase.org/). The MaizeGDB is the community database for biological information aboutZea mays, and includes genetic and genomic data sets and related information (http://www.maizegdb.org/) (Lawrence et al. 2004).

Genome-Wide Comparisons among Plants Species

With the completion of genome sequencing in a number of plant species, genome-scale comparative analyses can be used to produce and publish data sets that facilitate identification of conserved and/or characteristic properties among plant species.

Using modeled proteome data sets deduced from sequenced genomes in plants, several efforts have been completed to construct comprehensive gene families with the aim of establishing platforms to verify gene content and elucidating the process of gene duplication and functional diversification among species (Sterck et al. 2007). Comprehensive gene family data sets are usually produced by computational procedures including a step that conducts an all-against-all sequence similarity search and then a step for building clusters of protein families by methods such as Markov Clustering (MCL) or consideration of protein domain structures (Hulsen et al. 2006). The results of such studies can themselves yield databases that are useful for further phylogenetic studies (Horan et al. 2005, Conte et al. 2008, Wall et al. 2008).

Correlated gene arrangements among taxa along with chromosomal allocation, also known as synteny and collinearity, have become valuable frameworks for inference of shared ancestry of genes and for transfer of knowledge from a species to another related species (Tang et al. 2008a). The plant genome duplication database (PGDD) provides a data set of intragenome or cross-genome syntenic relationships identified throughout genome-sequenced plant species (http://chibba.agtec.uga.edu/duplication/) (Tang et al. 2008b).

Focused Database for Plant Genomics

Databases housing focused data sets together with rich annotations and well interrelated cross-references are also quite useful for the better understanding of focused issues in particular gene families and/or particular cellular processes.

Sequence-specific DNA-binding TFs are key molecular switches that control or influence many biological processes, such as development or responses to environmental changes. In plants, the genome-wide identification of repertories of genes encoding TFs of the Arabidopsis genome was reported first, and comparisons with other organisms revealed the properties of plant-specific TFs (Riechmann et al. 2000). In the past decade, with the availability of complete genome sequences, we have been able to compile catalogs describing the function and organization of TF regulatory systems in a number of organisms. There are many databases that provide data sets of genes putatively encoding TFs in many plant species; these are usually predictions based on computational methods such as sequence similarity search and/or hidden Markov model search of conserved DNA-binding domains (Table 5). Recently, further integration of data sets of TF-encoding genes has been performed, thus establishing an integrative, knowledge-based resource of TFs across related plant species in terms of comparative genomics of transcriptional regulatory networks. GRASSIUS provides the first step toward building a comprehensive platform for integration of information, tools and resources for comparative regulatory genomics across the grass species (Yilmaz et al. 2009). The Grass Transcription Factor Database (GrassTFDB) of GRASSIUS houses integrated information on MaizeTFDB, RiceTFDB, SorghumTFDB and CaneTFDB (http://grassius.org/grasstfdb.html). The LegumeTFDB provides predicted TF- encoding genes annotated in the genome sequences of three major legume species: soybean, L. japonicus and M. truncatula(http://legumetfdb.psc.riken.jp/). This database is an extended version of the SoybeanTFDB (http://soybeantfdb.psc.riken.jp/) and is aimed at integrating knowledge on legume TFs and providing a public resource for comparative genomics of the TFs of legumes, non-legume plants and other organisms (Mochida et al. 2009c, Mochida et al. 2010).

Table 5: Transcription factor database in plants

Database	URL	Species	References
RARTF	http://rarge.gsc.riken.jp/rartf/	Arabidopsis	Iida et al. (2005)
AGRIS, AtTFDB	http://arabidopsis.med.ohio-state.edu/AtTFDB/	Arabidopsis	Palaniswamy et al. (2006)
DATF	http://datf.cbi.pku.edu.cn/	Arabidopsis	Guo et al. (2005)
DRTF	http://drtf.cbi.pku.edu.cn/	Rice	Gao et al. (2006)
DPTF	http://dptf.cbi.pku.edu.cn/	Poplar	Zhu et al. (2007)
TOBFAC	http://compsysbio.achs.virginia.edu/tobfac/	Tobacco	Rushton et al. (2008)
SoybeanTFDB	http://soybeantfdb.psc.riken.jp/	Soybean	Mochida et al. (2009c)
PlantTFDB	http://planttfdb.cbi.pku.edu.cn/	22 plant species	Guo et al. (2008)
PlnTFDB	http://plntfdb.bio.uni-potsdam.de/v3.0/	20 plant species	Riano-Pachon et al. (2007)
GRASSIUS, GrassTFDB	http://grassius.org/grasstfdb.html	Maize, rice, sorghum, sugarcane	Yilmaz et al. (2009)

LegumeTFDB	http://legumetfdb.psc.riken.jp/	Soybean, Lotus japonicus, Medicago truncatula	Mochida et al. (2010)
DBD	http://dbd.mrc-lmb.cam.ac.uk/DBD/index.cgi?Home	>700 species	Wilson et al. (2008)

ACKNOWLEDGMENTS

The authors thank Takuhiro Yoshida, Lam-Son Phan Tran and Tetsuya Sakurai of RIKEN Plant Science Center for their valuable suggestions and comments.

REFERENCES

1. Adams MD, Soares MB, Kerlavage AR, Fields C, Venter JC. Rapid cDNA sequencing (expressed sequence tags) from a directionally cloned human infant brain cDNA library. Nat. Genet. 1993; 4: 373-380.
2. Agrawal GK, Yonekura M, Iwahashi Y, Iwahashi H, Rakwal R. System, trends and perspectives of proteomics in dicot plants. Part I: technologies in proteome establishment. J. Chromatogr. B 2005; 815:109-123.
3. Akhunov E, Nicolet C, Dvorak J. Single nucleotide polymorphism genotyping in polyploid wheat with the Illumina GoldenGate assay. Theor. Appl. Genet. 2009; 119:507-517.
4. Akiyama K, Chikayama E, Yuasa H, Shimada Y, Tohge T., Shinozaki K, et al. PRIMe: a Web site that assembles tools for metabolomics and transcriptomics. In Silico Biol. 2008; 8:339-345.
5. Alonso JM, Ecker JR. Moving forward in reverse: genetic technologies to enable genome-wide phenomic screens in Arabidopsis. Nat. Rev. Genet. 2006; 7:524-536.
6. A G, Jeong DH, Jung KH, Lee S. Reverse genetic approaches for functional genomics of rice. Plant Mol. Biol. 2005 ; 59 :111-123.

7. Andersen JS, Mann M. Organellar proteomics: turning inventories into insights. EMBO Rep. 2006; 7:874-879.
8. Andreeva A, Howorth D, Chandonia JM, Brenner SE, Hubbard TJ, Chothia C, et al. Data growth and its impact on the SCOP database: new developments. Nucleic Acids Res. 2008; 36: D419-D425.
9. Anisimov SV. Serial analysis of gene expression (SAGE): 13 years of application in research. Curr. Pharm. Biotechnol. 2008; 9:338-350.
10. Ansorge WJ. Next-generation DNA sequencing techniques. Nat. Biotechnol. 2009; 25:195-203.
11. Ashikari M, Sakakibara H, Lin S, Yamamoto T, Takashi T, Nishimura A, et al. Cytokinin oxidase regulates rice grain production. Science 2005; 309:741-745.
12. Baginsky S. Plant proteomics: concepts, applications, and novel strategies for data interpretation. Mass Spectrom. Rev. 2009; 28:93-120.
13. Barakat A, Wall PK, Diloreto S, Depamphilis CW, Carlson JE. Conservation and divergence of microRNAs in Populus. BMC Genomics 2007; 8:481.
14. Barkley NA, Wang ML. Application of TILLING and EcoTILLING as reverse genetic approaches to elucidate the function of genes in plants and animals. Curr. Genomics 2008; 9:212-226.
15. Barrett T, Troup DB, Wilhite SE, Ledoux P, Rudnev D, Evangelista C, et al. NCBI GEO: archive for high-throughput functional genomic data. Nucleic Acids Res. 2009; 37:D885-D890.
16. Benschop JJ, Mohammed S, O'Flaherty M, Heck AJ, Slijper M, Menke FL. Quantitative phosphoproteomics of early elicitor signaling in Arabidopsis. Mol. Cell. Proteomics 2007;6:1198-1214.
17. Bernardo AN, Bradbury PJ, Ma H, Hu S, Bowden RL, Buckler ES, et al. Discovery and mapping of single feature polymorphisms in wheat using Affymetrix arrays. BMC Genomics 2009;10:251.
18. Bertocchi F, Paci M. Applications of high-resolution solid-state NMR spectroscopy in food science. J. Agric. Food. Chem. 2008;56:9317-9327.

19. Bevan M. Objective: the complete sequence of a plant genome. Plant Cell 1997;9:476-478.
20. Bino RJ, Hall RD, Fiehn O, Kopka J, Saito K, Draper J, et al. Potential of metabolomics as a functional genomics tool. Trends Plant Sci. 2004;9:418-425.
21. Birnbaum K, Shasha DE, Wang JY, Jung JW, Lambert GM, Galbraith DW, et al. A gene expression map of the Arabidopsis root. Science 2003;302:1956-1960.
22. Blair MW, Torres MM, Giraldo MC, Pedraza F. Development and diversity of Andean-derived, gene-based microsatellites for common bean (Phaseolus vulgaris L.). BMC Plant Biol. 2009;9:100.
23. Bogdanov B, Smith RD. Proteomics by FTICR mass spectrometry: top down and bottom up. Mass Spectrom. Rev. 2005;24:168-200.
24. Boguski MS, Lowe TM, Tolstoshev CM. dbEST—database for 'expressed sequence tags'. Nat. Genet. 1993;4:332-333.
25. Brady SM, Orlando DA, Lee JY, Wang JY, Koch J, Dinneny JR, et al. A high-resolution root spatiotemporal map reveals dominant expression patterns. Science 2007;318:801-806.
26. Brady SM, Provart NJ. Web-queryable large-scale data sets for hypothesis generation in plant biology. Plant Cell 2009;21:1034-1051.
27. Brenner S, Johnson M, Bridgham J, Golda G, Lloyd DH, Johnson D, et al. Gene expression analysis by massively parallel signature sequencing (MPSS) on microbead arrays. Nat. Biotechnol. 2000;18:630-634.
28. Brown ME, Funk CC. Climate. Food security under climate change. Science 2008;319:580-581.
29. Caicedo AL, Williamson SH, Hernandez RD, Boyko A, Fledel-Alon A, York TL, et al. Genome-wide patterns of nucleotide polymorphism in domesticated rice. PLoS Genet. 2007;3:1745-1756.
30. Carollo V, Matthews DE, Lazo GR, Blake TK, Hummel DD, Lui N, et al. GrainGenes 2.0. an improved resource for the small-grains community. Plant Physiol. 2005;139:643-651.

31. Castellani F, van Rossum B, Diehl A, Schubert M, Rehbein K, Oschkinat H. Structure of a protein determined by solid-state magic-angle-spinning NMR spectroscopy. Nature 2002;420:98-102.

32. Chellappan P, Jin H. Discovery of plant microRNAs and short-interfering RNAs by deep parallel sequencing. Methods Mol. Biol. 2009;495:121-132.

33. Chen S, Harmon AC. Advances in plant proteomics. Proteomics 2006;6:5504-5516.

34. Cheng Y, Qi Y, Zhu Q, Chen X, Wang N, Zhao X, et al. New changes in the plasma-membrane-associated proteome of rice roots under salt stress. Proteomics 2009;9:3100-3114.

35. Chuang CF, Meyerowitz EM. Specific and heritable genetic interference by double-stranded RNA in Arabidopsis thaliana. Proc. Natl Acad. Sci. USA 2000;97:4985-4990.

36. Close TJ, Bhat PR, Lonardi S, Wu Y, Rostoks N, Ramsay L, et al. Development and implementation of high-throughput SNP genotyping in barley. BMC Genomics 2009;10:582.

37. Close TJ, Wanamaker S, Roose ML, Lyon M. HarvEST: an EST database and viewing software. Methods Mol. Biol. 2007;406:161-178.

38. Cogburn LA, Porter TE, Duclos MJ, Simon J, Burgess SC, Zhu JJ, et al. Functional genomics of the chicken—a model organism. Poult. Sci. 2007;86:2059-2094.

39. Comai L, Young K, Till BJ, Reynolds SH, Greene EA, Codomo CA, et al. Efficient discovery of DNA polymorphisms in natural populations by Ecotilling. Plant J. 2004;37:778-786.

40. Conte MG, Gaillard S, Lanau N, Rouard M, Perin C. GreenPhylDB: a database for plant comparative genomics. Nucleic Acids Res. 2008;36:D991-D998.

41. Crossa J, Burgueno J, Dreisigacker S, Vargas M, Herrera-Foessel SA, Lillemo M, et al. Association analysis of historical bread wheat germplasm using additive genetic covariance of relatives and population structure. Genetics 2007;177:1889-1913.

42. De Bodt S, Maere S, Van de Peer Y. Genome duplication and the origin of angiosperms. Trends Ecol. Evol. 2005;20:591-597.

43. de Hoon M, Hayashizaki Y. Deep cap analysis gene expression (CAGE): genome-wide identification of promoters, quantification of their expression, and network inference. Biotechniques 2008;44:627-628, 630, 632.

44. de la Fuente van Bentem S, Anrather D, Roitinger E, Djamei A, Hufnagl T, Barta A, et al. Phosphoproteomics reveals extensive in vivo phosphorylation of Arabidopsis proteins involved in RNA metabolism. Nucleic Acids Res. 2006;34:3267-3278.

45. Deleu W, Esteras C, Roig C, Gonzalez-To M, Fernandez-Silva I, Gonzalez-Ibeas DD, et al. A set of EST-SNPs for map saturation and cultivar identification in melon. BMC Plant Biol. 2009;9:90.

46. Dixon RA, Gang DR, Charlton AJ, Fiehn O, Kuiper HA, Reynolds TL, et al. Applications of metabolomics in agriculture. J. Agric. Food. Chem. 2006;54:8984-8994.

47. Donlin MJ. Using the Generic Genome Browser (GBrowse). Curr. Protoc. Bioinformatics 2007. Chapter 9: Unit 9 9.

48. Google Scholar

49. Droc G, Perin C, Fromentin S, Larmande P. OryGenesDB 2008 update: database interoperability for functional genomics of rice. Nucleic Acids Res. 2009;37:D992-D995.

50. Dunkley TP, Hester S, Shadforth IP, Runions J, Weimar T, Hanton SL, et al. Mapping the Arabidopsis organelle proteome. Proc. Natl Acad. Sci. USA 2006;103:6518-6523.

51. Duvick J, Fu A, Muppirala U, Sabharwal M, Wilkerson MD, Lawrence CJ, et al. PlantGDB: a resource for comparative plant genomics. Nucleic Acids Res. 2008;36:D959-D965.

52. Endo Y, Sawasaki T. High-throughput, genome-scale protein production method based on the wheat germ cell-free expression system. Biotechnol. Adv. 2003;21:695-713.

53. Endo Y, Sawasaki T. Cell-free expression systems for eukaryotic protein production. Curr. Opin. Biotechnol. 2006;17:373-380.

54. Ewing B, Hillier L, Wendl MC, Green P. Base-calling of automated sequencer traces using phred. I. Accuracy assessment. Genome Res. 1998;8:175-185.

55. Farrer RA, Kemen E, Jones JD, Studholme DJ. De novo assembly of the Pseudomonas syringae pv. syringae B728a genome using

Illumina/Solexa short sequence reads. FEMS Microbiol. Lett. 2009;291:103-111.

56. Fei Z, Tang X, Alba RM, White JA, Ronning CM, Martin GB, et al. Comprehensive EST analysis of tomato and comparative genomics of fruit ripening. Plant J. 2004;40:47-59.

57. Feltus FA, Wan J, Schulze SR, Estill JC, Jiang N, Paterson AH. An SNP resource for rice genetics and breeding based on subspecies indica and japonica genome alignments. Genome Res. 2004;14:1812-1819.

58. Feolo M, Helmberg W, Sherry S, Maglott DR. NCBI genetic resources supporting immunogenetic research. Rev. Immunogenet. 2000;2:461-467.

59. Fernie AR, Schauer N. Metabolomics-assisted breeding: a viable option for crop improvement? Trends Genet. 2009;25:39-48.

60. Fernie AR, Trethewey RN, Krotzky AJ, Willmitzer L. Metabolite profiling: from diagnostics to systems biology. Nat. Rev. Mol. Cell Biol. 2004;5:763-769.

61. Fiehn O, Kopka J, Dormann P, Altmann T, Trethewey RN, Willmitzer L. Metabolite profiling for plant functional genomics. Nat. Biotechnol. 2000;18:1157-1161.

62. Finn RD, Tate J, Mistry J, Coggill PC, Sammut SJ, Hotz HR, et al. The Pfam protein families database. Nucleic Acids Res. 2008;36:D281-D288.

63. Flicek P, Aken BL, Beal K, Ballester B, Caccamo M, Chen Y, et al. Ensembl 2008. Nucleic Acids Res. 2008;36:D707-D714.

64. Fox BG, Goulding C, Malkowski MG, Stewart L, Deacon A. Structural genomics: from genes to structures with valuable materials and many questions in between. Nat. Methods 2008;5:129-132.

65. Fu J, Keurentjes JJ, Bouwmeester H, America T, Verstappen FW, Ward JL, et al. System-wide molecular evidence for phenotypic buffering in Arabidopsis. Nat. Genet. 2009;41:166-167.

66. Fujita M, Mizukado S, Fujita Y, Ichikawa T, Nakazawa M, Seki M, et al. Identification of stress-tolerance-related transcription-factor genes via mini-scale Full-length cDNA Over-eXpressor (FOX) gene hunting system. Biochem. Biophys. Res. Commun. 2007;364:250-257.

67. Fukuchi S, Homma K, Sakamoto S, Sugawara H, Tateno Y, Gojobori T, et al. The GTOP database in 2009: updated content and novel features to expand and deepen insights into protein structures and functions. Nucleic Acids Res. 2009;37:D333-D337.

68. Futamura N, Totoki Y, Toyoda A, Igasaki T, Nanjo T, Seki M., et al. Characterization of expressed sequence tags from a full-length enriched cDNA library of Cryptomeria japonica male strobili. BMC Genomics 2008;9:383.

69. Gao G, Zhong Y, Guo A, Zhu Q, Tang W, Zheng W, et al. DRTF: a database of rice transcription factors. Bioinformatics 2006;22:1286-1287.

70. Gao J, Agrawal GK, Thelen JJ, Xu D. P3DB: a plant protein phosphorylation database. Nucleic Acids Res. 2009;37:D960-D962.

71. Goda H, Sasaki E, Akiyama K, Maruyama-Nakashita A, Nakabayashi K, Li W, et al. The AtGenExpress hormone and chemical treatment data set: experimental design, data evaluation, model data analysis and data access. Plant J. 2008;55:526-542.

72. Goff SA, Ricke D, Lan TH, Presting G, Wang R, Dunn M, et al. A draft sequence of the rice genome (Oryza sativa L. ssp. japonica). Science 2002;296:92-100.

73. Greene LH, Lewis TE, Addou S, Cuff A, Dallman T, Dibley M, et al. The CATH domain structure database: new protocols and classification levels give a more comprehensive resource for exploring evolution. Nucleic Acids Res. 2007;35:D291-D297.

74. Gstaiger M, Aebersold R. Applying mass spectrometry-based proteomics to genetics, genomics and network biology. Nat. Rev. Genet. 2009;10:617-627.

75. Guo A, He K, Liu D, Bai S, Gu X, Wei L, et al. DATF: a database of Arabidopsis transcription factors. Bioinformatics 2005;21:2568-2569.

76. Guo AY, Chen X, Gao G, Zhang H, Zhu QH, Liu XC, et al. PlantTFDB: a comprehensive plant transcription factor database. Nucleic Acids Res. 2008;36:D966-D969.

77. Guo HS, Fei JF, Xie Q, Chua NH. A chemical-regulated inducible RNAi system in plants. Plant J. 2003;34:383-392.

78. Haas BJ, Delcher AL, Wortman JR, Salzberg SL. DAGchainer: a tool for mining segmental genome duplications and synteny. Bioinformatics 2004;20:3643-3646.

79. Han B, Xue Y. Genome-wide intraspecific DNA-sequence variations in rice. Curr. Opin. Plant Biol. 2003;6:134-138.

80. Harbers M, Carninci P. Tag-based approaches for transcriptome research and genome annotation. Nat. Methods 2005;2:495-502.

81. Hashimoto S, Suzuki Y, Kasai Y, Morohoshi K, Yamada T, Sese J, et al. 5′-End SAGE for the analysis of transcriptional start sites. Nat. Biotechnol. 2004;22:1146-1149.

82. Hayashizaki Y. RIKEN mouse genome encyclopedia. Mech. Ageing Dev. 2003;124:93-102.

83. Heesacker A, Kishore VK, Gao W, Tang S, Kolkman JM, Gingle A, et al. SSRs and INDELs mined from the sunflower EST database: abundance, polymorphisms, and cross-taxa utility. Theor. Appl. Genet. 2008;117:1021-1029.

84. Hilson P, Small I, Kuiper MT. European consortia building integrated resources for Arabidopsis functional genomics. Curr. Opin. Plant Biol. 2003;6:426-429.

85. Hirai MY, Yano M, Goodenowe DB, Kanaya S, Kimura T, Awazuhara M, et al. Integration of transcriptomics and metabolomics for understanding of global responses to nutritional stresses in Arabidopsis thaliana. Proc. Natl Acad. Sci. USA 2004;101:10205-10210.

86. Hirano H, Islam N, Kawasaki H. Technical aspects of functional proteomics in plants. Phytochemistry 2004;65:1487-1498.

87. Hiratsu K, Matsui K, Koyama T, Ohme-Takagi M. Dominant repression of target genes by chimeric repressors that include the EAR motif, a repression domain, in Arabidopsis. Plant J. 2003;34:733-739.

88. Hishiki T, Kawamoto S, Morishita S, Okubo K. BodyMap: a human and mouse gene expression database. Nucleic Acids Res. 2000;28:136-138.

89. Horan K, Lauricha J, Bailey-Serres J, Raikhel N, Girke T. Genome cluster database. A sequence family analysis platform for Arabidopsis and rice. Plant Physiol. 2005;138:47-54.

90. Hori K, Sato K, Takeda K. Detection of seed dormancy QTL in multiple mapping populations derived from crosses involving novel barley germplasm. Theor. Appl. Genet. 2007;115:869-876.
91. Huang X, Feng Q, Qian Q, Zhao Q, Wang L, Wang A, et al. High-throughput genotyping by whole-genome resequencing. Genome Res. 2009;19:1068-1076.
92. Huang X, Madan A. CAP3: A DNA sequence assembly program. Genome Res. 1999;9:868-877.
93. Hulsen T, de Vlieg J, Groenen PM. PhyloPat: phylogenetic pattern analysis of eukaryotic genes. BMC Bioinformatics 2006;7:398.
94. Hunter S, Apweiler R, Attwood TK, Bairoch A, Bateman A, Binns D, et al. InterPro: the integrative protein signature database. Nucleic Acids Res. 2009;37:D211-D215.
95. Hyten DL, Song Q, Choi IY, Yoon MS, Specht JE, Matukumalli LK, et al. High-throughput genotyping with the GoldenGate assay in the complex genome of soybean. Theor. Appl. Genet. 2008;116:945-952.
96. Ichikawa T, Nakazawa M, Kawashima M, Iizumi H, Kuroda H, Kondou Y, et al. The FOX hunting system: an alternative gain-of-function gene hunting technique. Plant J. 2006;48:974-985.
97. Igawa T, Fujiwara M, Takahashi H, Sawasaki T, Endo Y, Seki M, et al. Isolation and identification of ubiquitin-related proteins from Arabidopsis seedlings. J. Exp. Bot. 2009;60:3067-3073.
98. Iida K, Seki M, Sakurai T, Satou M, Akiyama K, Toyoda T, et al. Genome-wide analysis of alternative pre-mRNA splicing in Arabidopsis thaliana based on full-length cDNA sequences. Nucleic Acids Res. 2004;32:5096-5103.
99. Iida K, Seki M, Sakurai T, Satou M, Akiyama K, Toyoda T, et al. RARTF: database and tools for complete sets of Arabidopsis transcription factors. DNA Res. 2005;12:247-256.
100. Iijima Y, Nakamura Y, Ogata Y, Tanaka K, Sakurai N, Suda K, et al. Metabolite annotations based on the integration of mass spectral information. Plant J. 2008;54:949-962.
101. Imanishi T, Itoh T, Suzuki Y, O'Donovan C, Fukuchi S, Koyanagi KO, et al. Integrative annotation of 21,037 human genes validated by full-length cDNA clones. PLoS Biol. 2004;2:e162.

102. International Rice Genome Sequencing Project. The map-based sequence of the rice genome. Nature 2005;436:793-800.

103. Islam N, Lonsdale M, Upadhyaya NM, Higgins TJ, Hirano H, Akhurst R. Protein extraction from mature rice leaves for two-dimensional gel electrophoresis and its application in proteome analysis. Proteomics 2004;4:1903-1908.

104. Itoh T, Tanaka T, Barrero RA, Yamasaki C, Fujii Y, Hilton PB, et al. Curated genome annotation of Oryza sativa ssp. japonica and comparative genome analysis with Arabidopsis thaliana. Genome Res. 2007;17:175-183.

105. Jaccoud D, Peng K, Feinstein D, Kilian A. Diversity arrays: a solid state technology for sequence information independent genotyping. Nucleic Acids Res. 2001;29:E25.

106. Jaillon O, Aury JM, Noel B, Policriti A, Clepet C, Casagrande A, et al. The grapevine genome sequence suggests ancestral hexaploidization in major angiosperm phyla. Nature 2007;449:463-467.

107. Jiao Y, Tausta SL, Gandotra N, Sun N, Liu T, Caly NK, et al. A transcriptome atlas of rice cell types uncovers cellular, functional and developmental hierarchies. Nat. Genet. 2009;41:258-263.

108. Jonsson P, Gullberg J, Nordstrom A, Kusano M, Kowalczyk M, Sjostrom M, et al. A strategy for identifying differences in large series of metabolomic samples analyzed by GC/MS. Anal. Chem. 2004;76:1738-1745.

109. Jorrin-Novo JV, Maldonado AM, Echevarria-Zomeno S, Valledor L, Castillejo MA, Curto M, et al. Plant proteomics update (2007–2008): second-generation proteomic techniques, an appropriate experimental design, and data analysis to fulfill MIAPE standards, increase plant proteome coverage and expand biological knowledge. J. Proteomics 2009;72:285-314.

110. Kakimoto T. CKI1, a histidine kinase homolog implicated in cytokinin signal transduction. Science 1996;274:982-985.

111. Kanehisa M, Araki M, Goto S, Hattori M, Hirakawa M, Itoh M, et al. KEGG for linking genomes to life and the environment. Nucleic Acids Res. 2008;36:D480-D484.

112. Kanehisa M, Goto S. KEGG: Kyoto encyclopedia of genes and genomes. Nucleic Acids Res. 2000;28:27-30.

113. Kantety RV, La Rota M, Matthews DE, Sorrells ME. Data mining for simple sequence repeats in expressed sequence tags from barley, maize, rice, sorghum and wheat. Plant Mol. Biol. 2002;48:501-510.

114. Katz A, Waridel P, Shevchenko A, Pick U. Salt-induced changes in the plasma membrane proteome of the halotolerant alga Dunaliella salina as revealed by blue native gel electrophoresis and nano-LC-MS/MS analysis. Mol. Cell. Proteomics 2007;6:1459-1472.

115. Kaur S, Cogan NO, Ye G, Baillie RC, Hand ML, Ling AE, et al. Genetic map construction and QTL mapping of resistance to blackleg (Leptosphaeria maculans) disease in Australian canola (Brassica napus L.) cultivars. Theor. Appl. Genet. 2009;120:71-83.

116. Kawamoto S, Yoshii J, Mizuno K, Ito K, Miyamoto Y, Ohnishi T, et al. BodyMap: a collection of 3′ ESTs for analysis of human gene expression information. Genome Res. 2000;10:1817-1827.

117. Kawaura K, Mochida K, Enju A, Totoki Y, Toyoda A, Sakaki Y, et al. Assessment of adaptive evolution between wheat and rice as deduced from full-length common wheat cDNA sequence data and expression patterns. BMC Genomics 2009;10:271.

118. Kawaura K, Mochida K, Yamazaki Y, Ogihara Y. Transcriptome analysis of salinity stress responses in common wheat using a 22k oligo-DNA microarray. Funct. Integr. Genomics 2006;6:132-142.

119. Kigawa T, Yabuki T, Yoshida Y, Tsutsui M, Ito Y, Shibata T, et al. Cell-free production and stable-isotope labeling of milligram quantities of proteins. FEBS Lett. 1999;442:15-19.

120. Kikuchi J, Shinozaki K, Hirayama T. Stable isotope labeling of Arabidopsis thaliana for an NMR-based metabolomics approach. Plant Cell Physiol. 2004;45:1099-1104.

121. Kikuchi S, Satoh K, Nagata T, Kawagashira N, Doi K, Kishimoto N, et al. Collection, mapping, and annotation of over 28,000 cDNA clones from japonica rice. Science 2003;301:376-379.

122. Kilian J, Whitehead D, Horak J, Wanke D, Weinl S, Batistic O, et al. The AtGenExpress global stress expression data set: protocols,

evaluation and model data analysis of UV-B light, drought and cold stress responses. Plant J. 2007;50:347-363.

123. Kolesnik T, Szeverenyi I, Bachmann D, Kumar CS, Jiang S. Establishing an efficient Ac/Ds tagging system in rice: large-scale analysis of Ds flanking sequences. Plant J. 2004;37:301-314.

124. Komatsu S. Rice proteome database: a step toward functional analysis of the rice genome. Plant Mol. Biol. 2005;59:179-190.

125. Kondou Y, Higuchi M, Takahashi S, Sakurai T, Ichikawa T, Kuroda H, et al. Systematic approaches to using the FOX hunting system to identify useful rice genes. Plant J. 2009;57:883-894.

126. Konishi S, Izawa T, Lin SY, Ebana K, Fukuta Y, Sasaki T, Yano M. An SNP caused loss of seed shattering during rice domestication. Science 2006;312:1392-1396.

127. Kopka J, Schauer N, Krueger S, Birkemeyer C, Usadel B, Burgmüller E, et al. GMD@CSB.DB: the Golm Metabolome Database. Bioinformatics 2005;21:1635-1638.

128. Kose F, Weckwerth W, Linke T, Fiehn O. Visualizing plant metabolomic correlation networks using clique-metabolite matrices. Bioinformatics 2001;17:1198-1208.

129. Kota R, Rudd S, Facius A, Kolesov G, Thiel T, Zhang H, et al. Snipping polymorphisms from large EST collections in barley (Hordeum vulgare L.). Mol. Genet. Genomics 2003;270:24-33.

130. Kota R, Varshney RK, Prasad M, Zhang H, Stein N, Graner A. EST-derived single nucleotide polymorphism markers for assembling genetic and physical maps of the barley genome. Funct. Integr. Genomics 2008;8:223-233.

131. Kota R, Varshney RK, Thiel T, Dehmer KJ, Graner A. Generation and comparison of EST-derived SSRs and SNPs in barley (Hordeum vulgare L.). Hereditas 2001;135:145-151.

132. Kouranov A, Xie L, de la Cruz J, Chen L, Westbrook J, Bourne PE, et al. The RCSB PDB information portal for structural genomics. Nucleic Acids Res. 2006;34:D302-305.

133. Krysan PJ, Young JC, Sussman MR. T-DNA as an insertional mutagen in Arabidopsis. Plant Cell 1999;11:2283-2290.

134. Kumar CS, Wing RA, Sundaresan V. Efficient insertional mutagenesis in rice using the maize En/Spm elements. Plant J. 2005;44:879-892.

135. Kurakawa T, Ueda N, Maekawa M, Kobayashi K, Kojima M, Nagato Y, et al. Direct control of shoot meristem activity by a cytokinin-activating enzyme. Nature 2007;445:652-655.

136. Kuromori T, Takahashi S, Kondou Y, Shinozaki K, Matsui M. Phenome analysis in plant species using loss-of-function and gain-of-function mutants. Plant Cell Physiol. 2009;50:1215-1231.

137. Kwon SJ, Choi EY, Choi YJ, Ahn JH, Park OK. Proteomics studies of post-translational modifications in plants. J. Exp. Bot. 2006;57:1547-1551.

138. Laubinger S, Zeller G, Henz SR, Sachsenberg T, Widmer CK, Naouar N, et al. At-TAX: a whole genome tiling array resource for developmental expression analysis and transcript identification in Arabidopsis thaliana. Genome Biol. 2008;9:R112.

139. Laurentin H, Ratzinger A, Karlovsky P. Relationship between metabolic and genomic diversity in sesame (Sesamum indicum L.). BMC Genomics 2008;9:250.

140. Lawrence CJ, Dong Q, Polacco ML, Seigfried TE, Brendel V. MaizeGDB, the community database for maize genetics and genomics. Nucleic Acids Res. 2004;32:D393-D397.

141. Lee Y, Tsai J, Sunkara S, Karamycheva S, Pertea G, Sultana R, et al. The TIGR Gene Indices: clustering and assembling EST and known genes and integration with eukaryotic genomes. Nucleic Acids Res. 2005;33:D71-D74.

142. Lein W, Usadel B, Stitt M, Reindl A, Ehrhardt T, Sonnewald U, et al. Large-scale phenotyping of transgenic tobacco plants (Nicotiana tabacum) to identify essential leaf functions. Plant Biotechnol. J. 2008;6:246-263.

143. Li F, Kitashiba H, Inaba K, Nishio T. A Brassica rapa linkage map of EST-based SNP markers for identification of candidate genes controlling flowering time and leaf morphological traits. DNA Res. 2009;16:311-323.

144. Liang C, Jaiswal P, Hebbard C, Avraham S, Buckler ES, Casstevens T, et al. Gramene: a growing plant comparative genomics resource. Nucleic Acids Res. 2008;36:D947-D953.

145. Lisec J, Meyer RC, Steinfath M, Redestig H, Becher M, Witucka-Wall H, et al. Identification of metabolic and biomass QTL

in Arabidopsis thaliana in a parallel analysis of RIL and IL populations. Plant J. 2008;53:960-972.

146. Liu X, Lu T, Yu S, Li Y, Huang Y, Huang T, et al. A collection of 10,096 indica rice full-length cDNAs reveals highly expressed sequence divergence between Oryza sativa indica and japonica subspecies. Plant Mol. Biol. 2007;65:403-415.

147. Long D, Martin M, Sundberg E, Swinburne J, Puangsomlee P, Coupland G. The maize transposable element system Ac/Ds as a mutagen in Arabidopsis: identification of an albino mutation induced by Ds insertion. Proc. Natl Acad. Sci. USA 1993;90:10370-10374.

148. Lu C, Kulkarni K, Souret FF, MuthuValliappan R, Tej SS, Poethig RS, et al. MicroRNAs and other small RNAs enriched in the Arabidopsis RNA-dependent RNA polymerase-2 mutant. Genome Res. 2006;16:1276-1288.

149. Lu R, Malcuit I, Moffett P, Ruiz MT, Peart J, Wu AJ, et al. High throughput virus-induced gene silencing implicates heat shock protein 90 in plant disease resistance. EMBO J. 2003a;22:5690-5699.

150. Lu R, Martin-Hernandez AM, Peart JR, Malcuit I, Baulcombe DC. Virus-induced gene silencing in plants. Methods 2003b;30:296-303.

151. Lu T, Yu S, Fan D, Mu J, Shangguan Y, Wang Z, et al. Collection and comparative analysis of 1888 full-length cDNAs from wild rice Oryza rufipogon Griff. W1943. DNA Res. 2008;15:285-295.

152. Ma JF, Tamai K, Yamaji N, Mitani N, Konishi S, Katsuhara M, et al. A silicon transporter in rice. Nature 2006;440:688-691.

153. Ma JF, Yamaji N, Mitani N, Tamai K, Konishi S, Fujiwara T, et al. An efflux transporter of silicon in rice. Nature 2007;448:209-212.

154. Mace ES, Rami JF, Bouchet S, Klein PE, Klein RR, Kilian A, et al. A consensus genetic map of sorghum that integrates multiple component maps and high-throughput Diversity Array Technology (DArT) markers. BMC Plant Biol. 2009;9:13.

155. Madin K, Sawasaki T, Ogasawara T, Endo Y. A highly efficient and robust cell-free protein synthesis system prepared from wheat

embryos: plants apparently contain a suicide system directed at ribosomes. Proc. Natl Acad. Sci. USA 2000;97:559-564.

156. Maeda N, Kasukawa T, Oyama R, Gough J, Frith M, Engström PG, et al. Transcript annotation in FANTOM3: mouse gene catalog based on physical cDNAs. PLoS Genet. 2006;2:e62.

157. Manzano C, Abraham Z, Lopez-Torrejon G, Del Pozo JC. Identification of ubiquitinated proteins in Arabidopsis. Plant Mol. Biol. 2008;68:145-158.

158. Maor R, Jones A, Nuhse TS, Studholme DJ, Peck SC, Shirasu K. Multidimensional protein identification technology (MudPIT) analysis of ubiquitinated proteins in plants. Mol. Cell. Proteomics 2007;6:601-610.

159. Margulies M, Egholm M, Altman WE, Attiya S, Bader JS, Bemben LA, et al. Genome sequencing in microfabricated high-density picolitre reactors. Nature 2005;437:376-380.

160. Marques MC, Alonso-Cantabrana H, Forment J, Arribas R, Alamar S, Conejero V, et al. A new set of ESTs and cDNA clones from full-length and normalized libraries for gene discovery and functional characterization in citrus. BMC Genomics 2009;10:428.

161. Maruyama K, Takeda M, Kidokoro S, Yamada K, Sakuma Y, Urano K, et al. Metabolic pathways involved in cold acclimation identified by integrated analysis of metabolites and transcripts regulated by DREB1A and DREB2A. Plant Physiol. 2009;150:1972-1980.

162. Masclaux F, Charpenteau M, Takahashi T, Pont-Lezica R, Galaud JP. Gene silencing using a heat-inducible RNAi system in Arabidopsis. Biochem. Biophys. Res. Commun. 2004;321:364-369.

163. Masoudi-Nejad A, Tonomura K, Kawashima S, Moriya Y, Suzuki M, Itoh M, et al. EGassembler: online bioinformatics service for large-scale processing, clustering and assembling ESTs and genomic DNA fragments. Nucleic Acids Res. 2006;34:W459-W462.

164. Mathews H, Clendennen SK, Caldwell CG, Liu XL, Connors K, Matheis N, et al. Activation tagging in tomato identifies a transcriptional regulator of anthocyanin biosynthesis, modification, and transport. Plant Cell 2003;15:1689-1703.

165. Matsuda F, Yonekura-Sakakibara K, Niida R, Kuromori T, Shinozaki K, Saito K. MS/MS spectral tag-based annotation of non-targeted profile of plant secondary metabolites. Plant J. 2009;57:555-577.

166. Matsui A, Ishida J, Morosawa T, Mochizuki Y, Kaminuma E, Endo TA, et al. Arabidopsis transcriptome analysis under drought, cold, high-salinity and ABA treatment conditions using a tiling array. Plant Cell Physiol. 2008;49:1135-1149.

167. Matsumura H, Kruger DH, Kahl G, Terauchi R. SuperSAGE: a modern platform for genome-wide quantitative transcript profiling. Curr. Pharm. Biotechnol. 2008;9:368-374.

168. Matsumura H, Reich S, Ito A, Saitoh H, Kamoun S, Winter P, et al. Gene expression analysis of plant host–pathogen interactions by SuperSAGE. Proc. Natl Acad. Sci. USA 2003;100:15718-15723.

169. Matsuzaki M, Misumi O, Shin IT, Maruyama S, Takahara M, Miyagishima SY, et al. Genome sequence of the ultrasmall unicellular red alga Cyanidioschyzon merolae 10D. Nature 2004;428:653-657.

170. McDermott A. Structure and dynamics of membrane proteins by magic angle spinning solid-state NMR. Annu. Rev. Biophys. 2009;38:385-403.

171. Mechin V, Balliau T, Chateau-Joubert S, Davanture M, Langella O, Negroni L, et al. A two-dimensional proteome map of maize endosperm. Phytochemistry 2004;65:1609-1618.

172. Mi H, Lazareva-Ulitsky B, Loo R, Kejariwal A, Vandergriff J, Rabkin S, et al. The PANTHER database of protein families, subfamilies, functions and pathways. Nucleic Acids Res. 2005;33:D284-D288.

173. Minami A, Fujiwara M, Furuto A, Fukao Y, Yamashita T, Kamo M, et al. Alterations in detergent-resistant plasma membrane microdomains in Arabidopsis thaliana during cold acclimation. Plant Cell Physiol. 2009;50:341-359.

174. Ming R, Hou S, Feng Y, Yu Q, Dionne-Laporte A, Saw JH, et al. The draft genome of the transgenic tropical fruit tree papaya (Carica papaya Linnaeus). Nature 2008;452:991-996.

175. Mitsuda N, Ohme-Takagi M. Functional analysis of transcription factors in Arabidopsis. Plant Cell Physiol. 2009;50:1232-1248.

176. Miyao A, Iwasaki Y, Kitano H, Itoh J, Maekawa M, Murata K, et al. A large-scale collection of phenotypic data describing an insertional mutant population to facilitate functional analysis of rice genes. Plant Mol. Biol. 2007;63:625-635.
177. Miyao A, Tanaka K, Murata K, Sawaki H, Takeda S, Abe K, et al. Target site specificity of the Tos17 retrotransposon shows a preference for insertion within genes and against insertion in retrotransposon-rich regions of the genome. Plant Cell 2003;15:1771-1780.
178. Mochida K, Furuta T, Ebana K, Shinozaki K, Kikuchi J. Correlation exploration of metabolic and genomic diversities in rice. BMC Genomics 2009a;10:568.
179. Mochida K, Kawaura K, Shimosaka E, Kawakami N, Shin-I T, Kohara Y, et al. Tissue expression map of a large number of expressed sequence tags and its application to in silico screening of stress response genes in common wheat. Mol. Gen. Genomics 2006;276:304-312.
180. Mochida K, Saisho D, Yoshida T, Sakurai T, Shinozaki K. TriMEDB: a database to integrate transcribed markers and facilitate genetic studies of the tribe Triticeae. BMC Plant Biol. 2008;8:72.
181. Mochida K, Yamazaki Y, Ogihara Y. Discrimination of homoeologous gene expression in hexaploid wheat by SNP analysis of contigs grouped from a large number of expressed sequence tags. Mol. Genet. Genomics 2003;270:371-377.
182. Mochida K, Yoshida T, Sakurai T, Ogihara Y, Shinozaki K. TriFLDB: a database of clustered full-length coding sequences from Triticeae with applications to comparative grass genomics. Plant Physiol. 2009b;150:1135-1146.
183. Mochida K, Yoshida T, Sakurai T, Yamaguchi-Shinozaki K, Shinozaki K, Tran LS. In silico analysis of transcription factor repertoire and prediction of stress responsive transcription factors in soybean. DNA Res. 2009c;16:353-369.
184. Mochida K, Yoshida T, Sakurai T, Yamaguchi-Shinozaki K, Shinozaki K, Tran LS. LegumeTFDB: an integrative database of Glycine max, Lotus japonicus and Medicago truncatula transcription factors. Bioinformatics 2010:26: 290-291.

185. Moco S, Bino RJ, Vorst O, Verhoeven HA, de Groot J, van Beek TA, et al. A liquid chromatography–mass spectrometry-based metabolome database for tomato. Plant Physiol. 2006;141:1205-1218.

186. Moran NA, McLaughlin HJ, Sorek R. The dynamics and time scale of ongoing genomic erosion in symbiotic bacteria. Science 2009;323:379-382.

187. Mori M, Tomita C, Sugimoto K, Hasegawa M, Hayashi N, Dubouzet JG, et al. Isolation and molecular characterization of a Spotted leaf 18 mutant by modified activation-tagging in rice. Plant Mol. Biol. 2007;63:847-860.

188. Morreel K, Goeminne G, Storme V, Sterck L, Ralph J, Coppieters W, et al. Genetical metabolomics of flavonoid biosynthesis in Populus: a case study. Plant J. 2006;47:224-237.

189. Mueller LA, Solow TH, Taylor N, Skwarecki B, Buels R, Binns J, et al. The SOL Genomics Network: a comparative resource for Solanaceae biology and beyond. Plant Physiol. 2005;138:1310-1317.

190. Mueller LA, Zhang P, Rhee SY. AraCyc: a biochemical pathway database for Arabidopsis. Plant Physiol. 2003;132:453-460.

191. Mulder NJ, Apweiler R. The InterPro database and tools for protein domain analysis. Curr. Protoc. Bioinformatics 2008. Chapter 2: Unit 2 7.

192. Nakamura H, Hakata M, Amano K, Miyao A, Toki N, Kajikawa M, et al. A genome-wide gain-of function analysis of rice genes using the FOX-hunting system. Plant Mol. Biol. 2007;65:357-371.

193. Nakano M, Nobuta K, Vemaraju K, Tej SS, Skogen JW, Meyers BC. Plant MPSS databases: signature-based transcriptional resources for analyses of mRNA and small RNA. Nucleic Acids Res. 2006;34:D731-D735.

194. Nanjo T, Futamura N, Nishiguchi M, Igasaki T, Shinozaki K, Shinohara K. Characterization of full-length enriched expressed sequence tags of stress-treated poplar leaves. Plant Cell Physiol. 2004;45:1738-1748.

195. Nanjo T, Sakurai T, Totoki Y, Toyoda A, Nishiguchi M, Kado D, et al. Functional annotation of 19,841 Populus nigra full-length enriched cDNA clones. BMC Genomics 2007;8:448.

196. Neale DB, Ingvarsson PK. Population, quantitative and comparative genomics of adaptation in forest trees. Curr. Opin. Plant Biol. 2008;11:149-155.

197. Newton RP, Brenton AG, Smith CJ, Dudley E. Plant proteome analysis by mass spectrometry: principles, problems, pitfalls and recent developments. Phytochemistry 2004;65:1449-1485.

198. Nishiyama T, Fujita T, Shin IT, Seki M, Nishide H, Uchiyama I, et al. Comparative genomics of Physcomitrella patens gametophytic transcriptome and Arabidopsis thaliana: implication for land plant evolution. Proc. Natl Acad. Sci. USA 2003;100:8007-8012.

199. Nobuta K, Lu C, Shrivastava R, Pillay M, De Paoli E, Accerbi M, et al. Distinct size distribution of endogeneous siRNAs in maize: evidence from deep sequencing in the mop1-1 mutant. Proc. Natl Acad. Sci. USA 2008;105:14958-14963.

200. Nobuta K, Venu RC, Lu C, Belo A, Vemaraju K, Kulkarni K, et al. An expression atlas of rice mRNAs and small RNAs. Nat. Biotechnol. 2007;25:473-477.

201. Nuhse TS, Bottrill AR, Jones AM, Peck SC. Quantitative phosphoproteomic analysis of plasma membrane proteins reveals regulatory mechanisms of plant innate immune responses. Plant J. 2007;51:931-940.

202. Nuhse TS, Stensballe A, Jensen ON, Peck SC. Large-scale analysis of in vivo phosphorylated membrane proteins by immobilized metal ion affinity chromatography and mass spectrometry. Mol. Cell. Proteomics 2003;2:1234-1243.

203. Nuhse TS, Stensballe A, Jensen ON, Peck SC. Phosphoproteomics of the Arabidopsis plasma membrane and a new phosphorylation site database. Plant Cell 2004;16:2394-2405.

204. Obayashi T, Hayashi S, Saeki M, Ohta H, Kinoshita K. ATTED-II provides coexpressed gene networks for Arabidopsis. Nucleic Acids Res. 2009;37:D987-D991.

205. Obayashi T, Kinoshita K, Nakai K, Shibaoka M, Hayashi S, Saeki M, et al. ATTED-II: a database of co-expressed genes and cis

elements for identifying co-regulated gene groups in Arabidopsis. Nucleic Acids Res. 2007;35:D863-D869.

206. Ogasawara O, Otsuji M, Watanabe K, Iizuka T, Tamura T, Hishiki T, et al. BodyMap-Xs: anatomical breakdown of 17 million animal ESTs for cross-species comparison of gene expression. Nucleic Acids Res. 2006;34:D628-D631.

207. Ogihara Y, Mochida K, Kawaura K, Murai K, Seki M, Kamiya A, et al. Construction of a full-length cDNA library from young spikelets of hexaploid wheat and its characterization by large-scale sequencing of expressed sequence tags. Genes Genet. Syst. 2004;79:227-232.

208. Ogihara Y, Mochida K, Nemoto Y, Murai K, Yamazaki Y, Shin IT, et al. Correlated clustering and virtual display of gene expression patterns in the wheat life cycle by large-scale statistical analyses of expressed sequence tags. Plant J. 2003;33:1001-1011.

209. Okazaki Y, Shimojima M, Sawada Y, Toyooka K, Narisawa T, Mochida K, et al. A chloroplastic UDP-glucose pyrophosphorylase from Arabidopsis is the committed enzyme for the first step of sulfolipid biosynthesis. Plant Cell 2009;21:892-909.

210. Oksman-Caldentey KM, Saito K. Integrating genomics and metabolomics for engineering plant metabolic pathways. Curr. Opin. Biotechnol. 2005;16:174-179.

211. Ossowski S, Schneeberger K, Clark RM, Lanz C, Warthmann N, Weigel D. Sequencing of natural strains of Arabidopsis thaliana with short reads. Genome Res. 2008;18:2024-2033.

212. Palaniswamy SK, James S, Sun H, Lamb RS, Davuluri RV, Grotewold E. AGRIS and AtRegNet. a platform to link cis-regulatory elements and transcription factors into regulatory networks. Plant Physiol. 2006;140:818-829.

213. Palczewski K, Kumasaka T, Hori T, Behnke CA, Motoshima H, Fox BA, et al. Crystal structure of rhodopsin: a G protein-coupled receptor. Science 2000;289:739-745.

214. Park PJ. ChIP-seq: advantages and challenges of a maturing technology. Nat. Rev. Genet. 2009;10:669-680.

215. Parkinson H, Kapushesky M, Shojatalab M, Abeygunawardena N, Coulson R, Farne A, et al. ArrayExpress—a public database

of microarray experiments and gene expression profiles. Nucleic Acids Res. 2007;35:D747-D750.

216. Paterson AH. Genomics of sorghum. Int. J. Plant Genomics 2008;2008:362451.

217. Paterson AH, Bowers JE, Bruggmann R, Dubchak I, Grimwood J, Gundlach H, et al. The Sorghum bicolor genome and the diversification of grasses. Nature 2009;457:551-556.

218. Peleg Z, Saranga Y, Suprunova T, Ronin Y, Roder MS, Kilian A, et al. High-density genetic map of durum wheat × wild emmer wheat based on SSR and DArT markers. Theor. Appl. Genet. 2008;117:103-115.

219. Qu S, Desai A, Wing R, Sundaresan V. A versatile transposon-based activation tag vector system for functional genomics in cereals and other monocot plants. Plant Physiol. 2008;146:189-199.

220. Ralph SG, Chun HJ, Cooper D, Kirkpatrick R, Kolosova N, Gunter L, et al. Analysis of 4,664 high-quality sequence-finished poplar full-length cDNA clones and their utility for the discovery of genes responding to insect feeding. BMC Genomics 2008a;9:57.

221. Ralph SG, Chun HJ, Kolosova N, Cooper D, Oddy C, Ritland CE, et al. A conifer genomics resource of 200,000 spruce (Picea spp.) ESTs and 6,464 high-quality, sequence-finished full-length cDNAs for Sitka spruce (Picea sitchensis). BMC Genomics 2008b;9:484.

222. Ramautar R, Somsen GW, de Jong GJ. CE-MS in metabolomics. Electrophoresis 2009;30:276-291.

223. Rensing SA, Lang D, Zimmer AD, Terry A, Salamov A, Shapiro H, et al. The Physcomitrella genome reveals evolutionary insights into the conquest of land by plants. Science 2008;319:64-69.

224. Riano-Pachon DM, Ruzicic S, Dreyer I, Mueller-Roeber B. PlnTFDB: an integrative plant transcription factor database. BMC Bioinformatics 2007;8:42.

225.

226. Riechmann JL, Heard J, Martin G, Reuber L, Jiang C, Keddie J, et al. Arabidopsis transcription factors: genome-wide comparative analysis among eukaryotes. Science 2000;290:2105-2110.

227. Roessner U, Luedemann A, Brust D, Fiehn O, Linke T, Willmitzer L, et al. Metabolic profiling allows comprehensive phenotyping of genetically or environmentally modified plant systems. Plant Cell 2001;13:11-29.

228. Rossignol M, Peltier JB, Mock HP, Matros A, Maldonado AM, Jorrin JV. Plant proteome analysis: a 2004–2006 update. Proteomics 2006;6:5529-5548.

229. Rostoks N, Borevitz JO, Hedley PE, Russell J, Mudie S, Morris J, et al. Single-feature polymorphism discovery in the barley transcriptome. Genome Biol. 2005;6:R54.

230. Rowe HC, Hansen BG, Halkier BA, Kliebenstein DJ. Biochemical networks and epistasis shape the Arabidopsis thaliana metabolome. Plant Cell 2008;20:1199-1216.

231. Ruiz-Ferrer V, Voinnet O. Roles of plant small RNAs in biotic stress responses. Annu. Rev. Plant Biol. 2009;60:485-510.

232. Rushton PJ, Bokowiec MT, Laudeman TW, Brannock JF, Chen X, Timko MP. TOBFAC: the database of tobacco transcription factors. BMC Bioinformatics 2008;9:53.

233. Saito K, Hirai MY, Yonekura-Sakakibara K. Decoding genes with coexpression networks and metabolomics—'majority report by precogs'. Trends Plant Sci. 2008;13:36-43.

234. Sakata K, Ohyanagi H, Nobori H, Nakamura T, Hashiguchi A, Nanjo Y, et al. Soybean proteome database: a data resource for plant differential omics. J. Proteome Res. 2009;8:3539-3548.

235. Sakurai T, Plata G, Rodriguez-Zapata F, Seki M, Salcedo A, Toyoda A, et al. Sequencing analysis of 20,000 full-length cDNA clones from cassava reveals lineage specific expansions in gene families related to stress response. BMC Plant Biol. 2007;7:66.

236. Sakurai T, Satou M, Akiyama K, Iida K, Seki M, Kuromori T, et al. RARGE: a large-scale database of RIKEN Arabidopsis resources ranging from transcriptome to phenome. Nucleic Acids Res. 2005;33:D647-D650.

237. Samatey FA, Imada K, Nagashima S, Vonderviszt F, Kumasaka T, Yamamoto M, et al. Structure of the bacterial flagellar protofilament and implications for a switch for supercoiling. Nature 2001;410:331-337.

238. Sato K, Shin IT, Seki M, Shinozaki K, Yoshida H, Takeda K, et al. Development of 5006 full-length cDNAs in barley: a tool for accessing cereal genomics resources. DNA Res. 2009;16:81-89.

239. Sato S, Nakamura Y, Kaneko T, Asamizu E, Kato T, Nakao M, et al. Genome structure of the legume, Lotus japonicus. DNA Res. 2008;15:227-239.

240. Sato S, Tabata S. Lotus japonicus as a platform for legume research. Curr. Opin. Plant Biol. 2006;9:128-132.

241. Schauer N, Semel Y, Balbo I, Steinfath M, Repsilber D, Selbig J, et al. Mode of inheritance of primary metabolic traits in tomato. Plant Cell 2008;20:509-523.

242. Schauer N, Semel Y, Roessner U, Gur A, Balbo I, Carrari F, et al. Comprehensive metabolic profiling and phenotyping of interspecific introgression lines for tomato improvement. Nat. Biotechnol. 2006;24:447-454.

243. Schena M, Shalon D, Davis RW, Brown PO. Quantitative monitoring of gene expression patterns with a complementary DNA microarray. Science 1995;270:467-470.

244. Schmelzle K, White FM. Phosphoproteomic approaches to elucidate cellular signaling networks. Curr. Opin. Biotechnol. 2006;17:406-414.

245. Schmutz J, Cannon SB, Schlueter J, Ma J, Mitros T, Nelson W, et al. Genome sequence of the palaeopolyploid soybean. Nature 2010;463:178-183.

246. Schnable PS, Ware D, Fulton RS, Stein JC, Wei F, Pasternak S, et al. The B73 maize genome: complexity, diversity, and dynamics. Science 2009;326:1112-1115.

247. Schneider A, Kirch T, Gigolashvili T, Mock HP, Sonnewald U, Simon R, et al. A transposon-based activation-tagging population in Arabidopsis thaliana (TAMARA) and its application in the identification of dominant developmental and metabolic mutations. FEBS Lett. 2005;579:4622-4628.

248. Schripsema J. Application of NMR in plant metabolomics: techniques, problems and prospects. Phytochem. Anal. 2009;21:14-21. Web of ScienceGoogle Scholar

249. Schwede T, Kopp J, Guex N, Peitsch MC. SWISS-MODEL: an automated protein homology-modeling server. Nucleic Acids Res. 2003;31:3381-3385.

250. Seki M, Narusaka M, Ishida J, Nanjo T, Fujita M, Oono Y, et al. Monitoring the expression profiles of 7000 Arabidopsis genes under drought, cold and high-salinity stresses using a full-length cDNA microarray. Plant J. 2002a;31:279-292.

251. Seki M, Narusaka M, Kamiya A, Ishida J, Satou M, Sakurai T, et al. Functional annotation of a full-length Arabidopsis cDNA collection. Science 2002b;296:141-145.

252. Seki M, Shinozaki K. Functional genomics using RIKEN Arabidopsis thaliana full-length cDNAs. J. Plant Res. 2009;122:355-366.

253. Sekiyama Y, Kikuchi J. Towards dynamic metabolic network measurements by multi-dimensional NMR-based fluxomics. Phytochemistry 2007;68:2320-2329.

254. Shen YJ, Jiang H, Jin JP, Zhang ZB, Xi B, He YY, et al. Development of genome-wide DNA polymorphism database for map-based cloning of rice genes. Plant Physiol. 2004;135:1198-1205.

255. Soderlund C, Descour A, Kudrna D, Bomhoff M, Boyd L, Currie J, et al. Sequencing, mapping, and analysis of 27,455 maize full-length cDNAs. PLoS Genet. 2009;5:e1000740.

256. Somerville C, Dangl J. Genomics. Plant biology in 2010. Science 2000;290:2077-2078.

257. Springer PS. Gene traps: tools for plant development and genomics. Plant Cell 2000;12:1007-1020.

258. Stanford WL, Cohn JB, Cordes SP. Gene-trap mutagenesis: past, present and beyond. Nat. Rev. Genet. 2001;2:756-768.

259. Sterck L, Rombauts S, Vandepoele K, Rouze P, Van de Peer Y. How many genes are there in plants (.... and why are they there)? Curr. Opin. Plant Biol. 2007;10:199-203.

260. Sterky F, Bhalerao RR, Unneberg P, Segerman B, Nilsson P, Brunner AM, et al. A Populus EST resource for plant functional genomics. Proc. Natl Acad. Sci. USA 2004;101:13951-13956.

261. Stevens RC, Yokoyama S, Wilson IA. Global efforts in structural genomics. Science 2001;294:89-92.

262. Sugiyama N, Nakagami H, Mochida K, Daudi A, Tomita M, Shirasu K, et al. Large-scale phosphorylation mapping reveals

the extent of tyrosine phosphorylation in Arabidopsis. Mol. Syst. Biol. 2008;4:193.

263. Sundaresan V, Springer P, Volpe T, Haward S, Jones JD, Dean C, et al. Patterns of gene action in plant development revealed by enhancer trap and gene trap transposable elements. Genes Dev. 1995;9:1797-1810.

264. Swarbreck D, Wilks C, Lamesch P, Berardini TZ, Garcia-Hernandez M, Foerster H, et al. The Arabidopsis Information Resource (TAIR): gene structure and function annotation. Nucleic Acids Res. 2008;36:D1009-D1014.

265. Taji T, Sakurai T, Mochida K, Ishiwata A, Kurotani A, Totoki Y, et al. Large-scale collection and annotation of full-length enriched cDNAs from a model halophyte, Thellungiella halophila. BMC Plant Biol. 2008;8:115.

266. Takeda S, Matsuoka M. Genetic approaches to crop improvement: responding to environmental and population changes. Nat. Rev. Genet. 2008;9:444-457.

267. Tanaka T, Antonio BA, Kikuchi S, Matsumoto T, Nagamura Y, Numa Y, et al. The Rice Annotation Project Database (RAP-DB): 2008 update. Nucleic Acids Res. 2008;36:D1028-D1033.

268. Tang H, Bowers JE, Wang X, Ming R, Alam M, Paterson AH. Synteny and collinearity in plant genomes. Science 2008a;320:486-488.

269. Tang H, Wang X, Bowers JE, Ming R, Alam M, Paterson AH. Unraveling ancient hexaploidy through multiply-aligned angiosperm gene maps. Genome Res. 2008b;18:1944-1954.

270. The Arabidopsis Genome Initiative. Analysis of the genome sequence of the flowering plant Arabidopsis thaliana. Nature 2000;408:796-815.

271. The International Brachypodium Initiative. Nature. Vol. 463. 2010. Genome sequencing and analysis of the model grass Brachypodium distachyon; p. 763-768.

272. Thimm O, Blasing O, Gibon Y, Nagel A, Meyer S, Kruger P, et al. MAPMAN: a user-driven tool to display genomics data sets onto diagrams of metabolic pathways and other biological processes. Plant J. 2004;37:914-939.

273. Tikunov Y, Lommen A, de Vos CH, Verhoeven HA, Bino RJ, Hall RD, et al. A novel approach for nontargeted data analysis for

metabolomics. Large-scale profiling of tomato fruit volatiles. Plant Physiol. 2005;139:1125-1137.

274. Till BJ, Colbert T, Codomo C, Enns L, Johnson J, Reynolds SH, et al. High-throughput TILLING for Arabidopsis. Methods Mol. Biol. 2006;323:127-135.

275. Till BJ, Reynolds SH, Weil C, Springer N, Burtner C, Young K, et al. Discovery of induced point mutations in maize genes by TILLING. BMC Plant Biol. 2004;4:12.

276. Tohge T, Nishiyama Y, Hirai MY, Yano M, Nakajima J, Awazuhara M, et al. Functional genomics by integrated analysis of metabolome and transcriptome of Arabidopsis plants over-expressing an MYB transcription factor. Plant J. 2005;42:218-235.

277. Tokimatsu T, Sakurai N, Suzuki H, Ohta H, Nishitani K, Koyama T, et al. KaPPA-view: a web-based analysis tool for integration of transcript and metabolite data on plant metabolic pathway maps. Plant Physiol. 2005;138:1289-1300.

278. Torada A, Koike M, Mochida K, Ogihara Y. SSR-based linkage map with new markers using an intraspecific population of common wheat. Theor. Appl. Genet. 2006;112:1042-1051.

279. Toyoda T, Wada A. Omic space: coordinate-based integration and analysis of genomic phenomic interactions. Bioinformatics 2004;20:1759-1765.

280. Toyoshima C, Nomura H, Sugita Y. Crystal structures of Ca2+-ATPase in various physiological states. Ann. NY Acad. Sci. 2003;986:1-8.

281. Trethewey RN. Metabolite profiling as an aid to metabolic engineering in plants. Curr. Opin. Plant Biol. 2004;7:196-201.

282. Turner WR, Oppenheimer M, Wilcove DS. A force to fight global warming. Nature 2009;462:278-279.

283. Tuskan GA, Difazio S, Jansson S, Bohlmann J, Grigoriev I, Helsten U, et al. The genome of black cottonwood, Populus trichocarpa (Torr. & Gray). Science 2006;313:1596-1604.

284. Tyler RC, Aceti DJ, Bingman CA, Cornilescu CC, Fox BG, Frederick RO, et al. Comparison of cell-based and cell-free protocols for producing target proteins from the Arabidopsis thaliana genome for structural studies. Proteins 2005;59:633-643.

285. Umezawa T, Sakurai T, Totoki Y, Toyoda A, Seki M, Ishiwata A, et al. Sequencing and analysis of approximately 40 000 soybean cDNA clones from a full-length-enriched cDNA library. DNA Res. 2008;15:333-346.

286. Urano K, Maruyama K, Ogata Y, Morishita Y, Takeda M, Sakurai N, et al. Characterization of the ABA-regulated global responses to dehydration in Arabidopsis by metabolomics. Plant J. 2009;57:1065-1078.

287. Varshney RK, Graner A, Sorrells ME. Genomics-assisted breeding for crop improvement. Trends Plant Sci. 2005;10:621-630.

288. Varshney RK, Nayak SN, May GD, Jackson SA. Next-generation sequencing technologies and their implications for crop genetics and breeding. Trends Biotechnol. 2009;27:522-530.

289. Velculescu VE, Zhang L, Vogelstein B, Kinzler KW. Serial analysis of gene expression. Science 1995;270:484-487.

290. von Roepenack-Lahaye E, Degenkolb T, Zerjeski M, Franz M, Roth U, Wessjohann L, et al. Profiling of Arabidopsis secondary metabolites by capillary liquid chromatography coupled to electrospray ionization quadrupole time-of-flight mass spectrometry. Plant Physiol. 2004; 134:548-559.

291. Wall PK, Leebens-Mack J, Muller KF, Field D, Altman NS, dePamphilis CW. PlantTribes: a gene and gene family resource for comparative genomics in plants. Nucleic Acids Res. 2008; 36:D970-D976.

292. Wang DK, Sun ZX, Tao YZ. Application of TILLING in plant improvement. Yi Chuan Xue Bao 2006; 33:957-964.

293. Ware D. Gramene: a resource for comparative grass genomics. Methods Mol. Biol. 2007; 406: 315-330.

294. Waterhouse PM, Graham MW, Wang MB. Virus resistance and gene silencing in plants can be induced by simultaneous expression of sense and antisense RNA. Proc. Natl Acad. Sci. USA 1998; 95: 13959-13964.

295. Weigel D, Ahn JH, Blazquez MA, Borevitz JO, Christensen SK, Fankhauser C, et al. Activation tagging in Arabidopsis. Plant Physiol. 2000; 122:1003-1013.

296. Weigel D, Mott R. The 1001 genomes project for Arabidopsis thaliana. Genome Biol. 2009; 10:107.

297. Wenzl P, Raman H, Wang J, Zhou M, Huttner E, Kilian A. A DArT platform for quantitative bulked segregant analysis. BMC Genomics 2007; 8:196.

298. Werner E, Heilier JF, Ducruix C, Ezan E, Junot C, Tabet JC. Mass spectrometry for the identification of the discriminating signals from metabolomics: current status and future trends. J. Chromatogr. B 2008; 871:143-163.

299. Wilson D, Charoensawan V, Kummerfeld SK, Teichmann SA. DBD—taxonomically broad transcription factor predictions: new content and functionality. Nucleic Acids Res. 2008; 36:D88-92.

300. Wilson D, Madera M, Vogel C, Chothia C, Gough J. The SUPERFAMILY database in 2007: families and functions. Nucleic Acids Res. 2007; 35: D308-D313.

301. Winter D, Vinegar B, Nahal H, Ammar R, Wilson GV, Provart NJ. An 'electronic fluorescent pictograph' browser for exploring and analyzing large-scale biological data sets. PLoS One 2007; 2:e718.

302. Yabuki T, Kigawa T, Dohmae N, Takio K, Terada T, Ito Y, et al. Dual amino acid-selective and site-directed stable-isotope labeling of the human c-Ha-Ras protein by cell-free synthesis. J. Biomol. NMR 1998; 11:295-306.

303. Yamamoto YY, Obokata J. ppdb: a plant promoter database. Nucleic Acids Res. 2008; 36:D977-D981.

304. Yamamoto YY, Yoshitsugu T, Sakurai T, Seki M, Shinozaki K, Obokata J. Heterogeneity of Arabidopsis core promoters revealed by high-density TSS analysis. Plant J. 2009; 60:350-362.

305. Yamasaki C, Murakami K, Fujii Y, Sato Y, Harada E, Takeda J, et al. The H-Invitational Database (H-InvDB), a comprehensive annotation resource for human genes and transcripts. Nucleic Acids Res. 2008a; 36:D793-D799.

306. Yamasaki K, Kigawa T, Inoue M, Tateno M, Yamasaki T, Yabuki T, et al. Solution structure of the B3 DNA binding domain of the Arabidopsis cold-responsive transcription factor RAV1. Plant Cell 2004; 16:3448-3459.

307. Yamasaki K, Kigawa T, Inoue M, Watanabe S, Tateno M, Seki M, et al. Structures and evolutionary origins of plant-specific

transcription factor DNA-binding domains. Plant Physiol. Biochem. 2008b; 46:394-401.

308. Yates JR, Ruse CI, Nakorchevsky A. Proteomics by mass spectrometry: approaches, advances, and applications. Annu. Rev. Biomed. Eng. 2009; 11:49-79.

309. Yilmaz A, Nishiyama M.Y. Jr., Fuentes BG, Souza GM, Janies D, Gray, J., et al. GRASSIUS: a platform for comparative regulatory genomics across the grasses. Plant Physiol. 2009; 149:171-180.

310. Yokoyama S. Protein expression systems for structural genomics and proteomics. Curr. Opin. Chem. Biol. 2003; 7:39-43.

311. Yokoyama S, Hirota H, Kigawa T, Yabuki T, Shirouzu M, Terada T, et al. Structural genomics projects in Japan. Nat. Struct. Biol. 2000;7(Suppl):943-945

312. Yonekura-Sakakibara K, Tohge T, Matsuda F, Nakabayashi R, Takayama H, Niida R, et al. Comprehensive flavonol profiling and transcriptome coexpression analysis leading to decoding gene–metabolite correlations in Arabidopsis. Plant Cell 2008; 20:2160-2176.

313. Yu J, Hu S, Wang J, Wong GK, Li S, Liu B, et al. A draft sequence of the rice genome (Oryza sativa L. ssp. indica). Science 2002; 296:79-92.

314. Zeller G, Henz SR, Widmer CK, Sachsenberg T, Ratsch G, Weigel D, et al. Stress-induced changes in the Arabidopsis thaliana transcriptome analyzed using whole-genome tiling arrays. Plant J. 2009;58: 1068-1082.

315. Zhang H, Sreenivasulu N, Weschke W, Stein N, Rudd S, Radchuk V, et al. Large-scale analysis of the barley transcriptome based on expressed sequence tags. Plant J. 2004;40:276-290.

316. Zhang J, Xu Y, Huan Q, Chong K. Deep sequencing of Brachypodium small RNAs at the global genome level identifies microRNAs involved in cold stress response. BMC Genomics 2009a;10:449

317. Zhang X, Yazaki J, Sundaresan A, Cokus S, Chan SW, Chen H, et al. Genome-wide high-resolution mapping and functional analysis of DNA methylation in arabidopsis. Cell 2006;126:1189-1201.

318. Zhang Y. Progress and challenges in protein structure prediction. Curr. Opin. Struct. Biol. 2008;18:342-348.

319. Zhang Y. I-TASSER: fully automated protein structure prediction in CASP8. Proteins 2009a;77(Suppl 9):100-113.

320. Zhang Y. Protein structure prediction: when is it useful? Curr. Opin. Struct. Biol. 2009b;19:145-155.

321. Zhang Z, Yu J, Li D, Liu F, Zhou X, Wang T, Ling Y, Su Z. PMRD: plant microRNA database. Nucleic Acids Res. 2009b.

322. Zheng Y, Ren N, Wang H, Stromberg AJ, Perry SE. Global identification of targets of the Arabidopsis MADS domain protein AGAMOUS-Like15. Plant Cell 2009;21:2563-2577.

323. Zhu QH, Guo AY, Gao G, Zhong YF, Xu M, Huang M, et al. DPTF: a database of poplar transcription factors. Bioinformatics 2007;23:1307-1308.

324. Zimmermann P, Hirsch-Hoffmann M, Hennig L, Gruissem W. GENEVESTIGATOR. Arabidopsis microarray database and analysis toolbox. Plant Physiol. 2004; 136: 2621-2632.

325. Zubko E, Adams CJ, Machaekova I, Malbeck J, Scollan C, Meyer P. Activation tagging identifies a gene from Petunia hybrida responsible for the production of active cytokinins in plants. Plant J. 2002; 29: 797-808.

Citations

CHAPTER 1

Liyuan Zhoua, , Ruiyuan Lib, Longjiang Fana, Yuewei Shib, Zhihong Wangb, Shengdon Xieb, Yijie Guia, Xueliang Renb, and Jun Zhua, Mapping epistasis and environment × QTX interaction based on four -omics genotypes for the detected QTX loci controlling complex traits in tobacco, doi:10.1016/j.cj.2013.09.001.

CHAPTER 2

Antonio J. Hall and Richard A. Richards, Prognosis for Genetic Improvement of Yield Potential and Water-Limited Yield of Major Grain Crops, doi:10.1016/j.fcr.2012.05.014.

CHAPTER 3

Roel C. Rabara, Marilyn C. Ferrer, Celia L. Diaz, Ma. Cristina V. Newingham, and Gabriel O. Romero, Phenotypic Diversity of Farmers' Traditional Rice Varieties in the Philippines, doi: 10.3390/agronomy4020217.

CHAPTER 4

Barend Juan Vorster, Urte Schlüter, Magdeleen du Plessis, Stefan van Wyk, Matome Eugene Makgopa, Ignatious Ncube, Marian Dorcas Quain, Karl Kunert, and Christine Helen Foyer, The Cysteine Protease–Cysteine Protease Inhibitor System Explored in Soybean Nodule Development, doi:10.3390/agronomy3030550.

CHAPTER 5

Robert Redden, New Approaches for Crop Genetic Adaptation to the Abiotic Stresses Predicted with Climate Change, doi: 10.3390/agronomy3020419.

CHAPTER 6

Catherine Tétard-Jones, Martin G. Edwards, Leonidas Rempelos, Angharad M.R. Gatehouse, Mick Eyre, Stephen J. Wilcockson, and Carlo Leifert, Effects of Previous Crop Management, Fertilization Regime and Water Supply on Potato Tuber Proteome and Yield, doi:10.3390/agronomy3010059.

CHAPTER 7

Farit M. Afendia, Naoaki Ono, Yukiko Nakamura, Kensuke Nakamura, Latifah K. Darusman, Nelson Kibinge, Aki Hirai Morita, Ken Tanaka, Hisayuki Horai, Md. Altaf-Ul-Amin, and Shigehiko Kanayaa, Data

Mining Methods for Omics and Knowledge of Crude Medicinal Plants toward Big Data Biology, doi:10.5936/csbj.201301010.

CHAPTER 8

Keiichi Mochida, and Kazuo Shinozaki, Genomics and Bioinformatics Resources for Crop Improvement, doi: 10.1093/pcp/pcq027.

Index

W

Y